THE ROAD TO SUCCESS

FOR

HIGH TECH CAREERS

Jerry Heines, PhD

ISBN-13: 978-1522976837

ISBN-10: 1522976833

To those who are beginning a search for a High Tech career,
And have received job offers to ponder,
Remember the immortal words of Yogi Berra,

"When you reach a fork in the road....Take it!"

ABOUT THE AUTHOR

Jerry received his PhD in Theoretical High Energy Particle Physics from the University of Connecticut. His pioneering work with his advisor, Dr. Munir Islam, remains today the most accepted model of the nucleon for physicists. Jerry's career has been focused on developing products for both Sonar and Radar applications used by the US Military and our NATO allies. In addition to studying science, Jerry is a public speaker and travels globally speaking about the latest technologies and Physics. He is often invited to speak about the most esoteric branches of physics to non-technical audiences as he delights in helping them appreciate and understand key concepts. A phrase passed down through several generations of Physicists to Jerry remains his focal point when speaking, "If you can not describe the essence of Physics so that your Grandmother can understand, than you yourself do not understand". In addition to studying physics and public speaking Jerry enjoys writing as evidenced by his recent publication "The Other Side of Darkness" which is a SciFi story followed by the actual Physics used to create the story. You can find Jerry's books on Amazon and at major book stores. Today he lives with his wife in sunny Florida.

For further information, please contact Jerry through Twitter @DrJerry2014 or via his LinkedIn account.

TABLE OF CONTENTS

LIST OF FIGURES

ACKNOWLEDGEMENTS

I am very grateful to the AIP Statistical Research Center (www.aip.org/statistics) for their generous support of this project through the use of the data on employment breakdown as well as salary data. Also, I would like to thank in no particular order, the reviewers of this manuscript. Mr. Peter L. Delos RF Engineer at Lockheed Martin, Dr. Chuck Kaufman Professor of Physics at the University of Rhode Island, Dr. Gary DeLeo Professor of Physics at Lehigh University, Dr. Art Hebard Professor of Physics at University of Florida, Dr. Xiaochang Miao Sr. Device Engineer, SanDisk. Your feedback and support on this project is beyond words but despite this meek apology I would like to express my heartfelt thanks to all of you to endure my badgering and questions. I've worked for many years with some of you and others I have had a long enjoyable professional relationship. But one commonality that you all share besides Physics, Engineering and Science is that I've learned from everyone…so thank you as your feedback has been a tremendous help to me personally and to make this project successful. And to my publisher at CreateSpace who has been extremely helpful and always liberal with time to help prepare this manuscript for publication.

Department of Physics
Lewis Laboratory
16 Memorial Drive East
Bethlehem, PA 18015-3182
(610) 758-3930 Fax (610) 758-3730
http://www.physics.lehigh.edu

To Whom it May Concern:

It is my pleasure to provide the following endorsement...

The Road to Success for High Tech Careers is a must-read for those with higher-education degrees in the physical sciences or engineering who aspire to attain and succeed in industrial careers. Dr. Heines combines a broad range of personal experiences with a narrative style filled with lighthearted humor and serious warnings. This book will help one move along the road to success even for those graduating students with advisors well-connected with industry (many are not) and those currently in industry who wish to thrive and grow in the corporate environment. *Gary G. DeLeo, Professor of Physics, Lehigh University.*

Sincerely yours,

Gary G. DeLeo
Professor of Physics
Lehigh University

9

UF | UNIVERSITY *of*
FLORIDA

College of Liberal Arts and Sciences
Department of Physics

PO Box 118440
Gainesville, FL 32611-8440
352-392-8842
352-392-3591 Fax

May 26, 2015

Re: Recommendation for "The Road to Success for High Tech Careers" by Jerry Heines

To whom it may concern:

The process of introducing a broad spectrum of students to the excitement of science and then concentrating on a select few to pursue a PhD degree in STEM areas is not very satisfying without meaningful closure for all students at the end stages of degree conferral. Will the affected students enjoy satisfying careers and be able to make meaningful contributions in their chosen areas of expertise? What are the career alternatives and what is the most efficient way of matching students to jobs and expectations to outcomes? This manuscript, "The Road to Success for High Tech Careers" by Jerry Heines provides answers to these questions and is an essential resource for students interested in pursuing high tech jobs in the corporate world. The author's experience as a PhD physicist combined with entrepreneurial immersion in the corporate world provides a perspective that cannot be found at career fairs or in the professional development programs of technical societies/organizations. Students and their respective advisors will benefit from the advice given on currently available job opportunities, life styles inside the corporate world, interview and resume writing tips, career advancement, and the pros and cons of small versus large corporations. I highly recommend this book to students who, with their degrees in hand, are seeking employment in high tech careers.

Sincerely,

Arthur F. Hebard
Distinguished Professor of Physics
Physics Department,
2001 Museum Road
P.O. Box 118440 Gainesville
FL 32611-8440
Email: afh@phys.ufl.edu
Phone: 352 392-8842
Fax: 352 392-3591

The Foundation for The Gator Nation
An Equal Opportunity Institution

10

6/30/2015

Delos, Peter L
Lockheed Martin

199 Borton Landing Rd
Moorestown, NJ 08057

To Whom It May Concern

For young professionals finishing college and entering the workplace, the transition can be a daunting experience. There is much activity in job fairs, but limited practical guides about behaviors needed for a successful transition. The recent publication "The Road to Success for High Tech Careers" by Jerry Heines fulfills a needed gap and should be a booklet in every University job resource room. Jerry has summarized his experience, consulted many people in industry, incorporated many comments, and ultimately included the experience of many professionals in his work. The culmination of these experiences documented in a concise manner will be a valuable resource to young professionals for many years to come.

Peter Delos
RF Engineer
Lockheed Martin

1.   INTRODUCTION

If you are reading this you are probably close to graduation and about to consider the next step in life. Congratulations first before we begin to explore some concepts that will help you in moving forward. It is a great achievement to finish a technical program which could be in the field of Electrical Engineering, Physics, Mechanical Engineering, Chemistry, Biology, Mathematics/Statistics, Chemistry, Aerospace Engineering, Geology, etc. These curriculums are not for the faint of heart as they take a great deal of commitment to complete. This is important to recognize as it helps to build self-confidence. And as you come to the pinnacle of one area you step into another, only to find that as you get to the top of one achievement it means you then enter at the bottom of another. This is life and it will happen many, many times throughout your career. It is fun. It is exciting. And it is rewarding. Look back from where you came and then look forward to where you are going. You get to make the choices; no one else makes them for you but making informed choices is the key. If you are a student you are very aware that to take that first step from school to a job is an important change happening in your life and to help offer some guidance on that is the focus of this book.

Or perhaps you are considering a change in career. Today the boundaries between technologies and science are rather gray as there is considerable overlap. And gone are the days where most of us will spend our entire career working in just one area. Just as when you started working right out of school you will find that what you learned from one area along your career path can be applied to another aspect of your professional employment. You apply your existing knowledge to your current job, add new skills and continue to expand. And perhaps you wonder what might be exciting to consider for a change in career and what might be the most effective method to pursue this new dream?

We often find that what we trained for in school or learned while at our current job is not exactly what we end up doing in our career. So you may even find that you question yourself if what you are doing is the right job for you. And why did you go to school in the first place if it is not in an area that you studied? This is always a good question to ask yourself as you advance in your career path as it helps you to search your soul. By reaching deep within yourself to evaluate different options that may come your way you will find the right path. One element I found early on was not to chase the big bucks as the main driver in your decision. The thrill of the check wears off fast after a few weeks of work in an area that you find you do not enjoy. And jumping quickly from job to job does not look good to a prospective employer. So it comes back to making informed decisions.

So why go to school if you do not end up working in an area that you studied? You now have refined skills for problems solving capability. It is that simple....in school you are thrown many different problems to solve and this develops the mental capabilities needed to identify, define and solve a variety of issues in a logical, progressive manner. And how do you obtain these attributes? It all comes from learning research methods used in the past as well as the present, analytical capabilities, writing reports, presenting results, approaches to problem solving and wide array of other talents. Another area that school offers is social. Working together with a wide range of colleagues is an important skill because in industry you will be a part of a team. You need to learn how to integrate into a team productively so that you feel you contribute and the team moves forward on the assigned project. Everyone is happy including the boss as the bottom line and keeping valued customers happy is of the utmost importance for the company's success and longevity.

But first, you need to get a job. One item before we begin however; you will note that I use 'his' and 'her' interchangeably in this book as it can be either sex that you will interact with in today's world – and I think that is absolutely wonderful given how it used to be when I first started in the male dominated world of business! I have seen many positives that have come from women assuming more high level roles and it is a pleasure to see that it is only getting better! So if you are ready, here we go …..

2.   A CAREER PATH EXAMPLE

One surprise I had as I was finishing up my doctorate in Physics was that none of the faculty had any industrial experience. I was offered a number of post-docs from different universities, but did not want to get into that cycle as there was no guarantee to land a permanent position. So that eliminated an academic career as a professor (which I was really focused on when I started grad school) and so I turned my attention to industry. As there were no faculty at the university (and for the most part, this is true even today in that there are very few professors connected to industry) who could make introductions for industrial positions, I was left in a real quandary! How does one with a PhD in Theoretical High Energy Particle Physics get a job in industry? Remember the soul searching mentioned above? Well after doing a great deal of this I found that the connections to industry I could offer stemmed from my analytical or math and computer skills. In addition problem solving was another area to connect to industry. I developed the attitude right at the beginning that I would look for ways to find industrial problems (whatever they might be, as I had no insight into this area – there was no internet when I was starting) exciting and challenging so they would satisfy my desire to enjoy my profession. So starting with a positive mind set and a means to connect my education to whatever was out there, I began to seek interviews.

Personally, examples help me understand the methodologies available for problem solving. In that vein, here is a brief outline of my career highlights and key elements in my decision making. Some of my decisions were voluntary, some were necessary as the market 'tanked' and people were laid off. All this may happen during your career so just expect it and roll with it when it happens. Keep a strong mind set, positive outlook and continue looking for new ways to stay challenged, productive and enjoy what you do. And never give up!

<u>Highlights</u>:

1. PhD in Theoretical High Energy Particle Physics
   a. Decided math skills were portable to numerous industrial applications
   b. During interviews found there were many problems that were interesting to study and solve in industry.
2. Worked as an Applied Physicist analyzing Navy SONAR issues
   a. Worked in a Navy Think Tank
   b. Studied ways to optimize various Navy actions to Cold War threats
   c. Developed an analytical scattering model of sound from fluid-loaded, frame-stiffened cylinders (like submarines, e.g.)
   d. DoD funding significantly reduced → changed careers
3. Radar and EW (Electronic Warfare) hardware development
   a. Identified the synergy between Sonar and Radar signal processing as a selling tool for me to change markets and made this connection so it was clear during interviews.
   b. Hired into the electronic realm of military industry
   c. Studied second order physics effects in crystals used in signal processing
   d. Developed software tools for hardware designers
   e. Automated test stations in the labs to multitask and thermally cycle products
   f. Customers became more informed and asked more detailed questions.
   g. Began to visit customers with Sales teams to understand their future needs and explain technical details.
4. Moved to Engineering Management
   a. More customer visits forced me to move off the bench to support their needs.

b. Tasked with hiring engineers and physicists
c. Program management responsibilities, less time on-the-bench in product development and R&D
d. Customer interfacing to answer technical questions increased
e. Found a major weakness: customers and suppliers did not speak the same language as engineers. This caused major issues with customers, so changed careers to address this need.
5. Technical Marketing
   a. Focused on speaking about technologies that were available to address various threats.
   b. Traveled globally to all NATO friendly countries speaking on high tech solutions
   c. Helped develop corporate road maps for future products.

So these are my career highlights and as you read them you will see the key reasons for change which occurred along my path. In all cases, I was either forced to make a change due to market funding or did so to fill a void and take on new challenges as I grew professionally. I hope it helps quiet some concerns that you may have. Keep an open mind and look for the interesting aspects of whatever you do and you will find many rewarding opportunities in industry. For example, there are so many ways that one can find useful applications for Physics as noted in Figure 1, Newton's Law of Attraction.

Figure 1. Newton's Law of Attraction (Ref. Fig. 1)

3. WHAT'S OUT THERE (SOME EXAMPLES)

It is estimated that over one million jobs will be opened in STEM related (Science, Technology, Engineering, Math) areas by the year 2018. This explosion is happening now in fact. This section begins to look at areas which exhibit a high growth rate that is expected to be sustained in the coming years. The examples provided in this section are not at all comprehensive but serve more as a means for you to see what applications are being considered important in today's world. Industry is profit driven – no surprise to anyone with this brilliant insight. So don't be surprised to find that the mood and funding within a company changes with the financial climate. Most of the time corporations can see changes that are necessary in the market place in order for them to grow and remain prosperous. This is the challenge of the marketing departments, namely, to watch for and advise on:

1. direction for upcoming product needs,
2. expected prices that can be supported on the anticipated future program,
3. timing for these new products for market entry,
4. the technical performance that the products will need to meet and
5. the quantities expected in production.

These five elements listed above are important as they collectively constitute the company's road map for future growth. Technical marketing is fast, dynamic and carries a great deal of responsibility. If you get it wrong you would be out of a job as the company's finances could seriously suffer. It is a highly visible job to upper managers and company owners. But in this role, you see the big picture of what is happening in the world's market which makes it very intriguing and exciting. However, it takes years of experience before any company will trust one with this responsibility, as you can imagine.

So with the above brief introduction, let's look at a list of possible companies and professions. The big defense OEMs (Original Equipment Manufacturers) like Lockheed, Raytheon, etc. are so diverse that to see what they do and the locations (they are global) where the work is being done requires one look on their website. It is impossible to really do justice to them in this treatise.

Since the biggest employer today is still the Defense Sector we will start there and then look at other commercial sectors. Here are some examples of market areas which are growing today and will continue to hire technical personnel. Some examples of the different programs are given in parenthesis.

1. Department of Defense (DoD):
    a. Radar
        i. Ground based (incoming low flying aircraft or missile attacks)
        ii. Airborne (coastline for surveillance, downward looking)
        iii. Ship borne (often look for periscopes of small threat subs)
        iv. Space (high resolution imaging)
    b. Electronic Warfare (EW)
        i. Passive (namely one just looks for specific E&M spectra of the enemy)
        ii. Tactics may include directing energy at specific locations to control the enemy
        iii. Target humans, communication links, radar and other assets.
        iv. Includes IR, visible, UV and other frequency bands
    c. Missiles
        i. SAM (Surface Air Missiles), SSM (Surface Surface Missiles), ASM (Air Surface Missiles), AAM (Air Air Missiles)
        ii. Inertia (namely their mass is the payload)
    d. Unmanned Vehicles
        i. Surveillance
        ii. Weapon use
    e. Sonar
        i. Variety of underwater applications.

    f. Think Tank
        i. Battlefield theatre group
            1. Analyze future threat scenarios that could be played out (usually the Pentagon provides these or high ranking military officials) and optimize the performance
                a. usually the DoD funds this directly
                b. Optimize statistical searches
                c. Analyze battlefield scenarios
                d. Coordination of air/land/sea/space
                e. Counter Intelligence

2. Space studies of importance
    a. Key design issues being studied include smaller packages, higher performance and low cost
    b. Boundary layers (shock waves, E&M transmission and reception modeling)
    c. Structural (g-sensitivity, light materials but strong [using spider webs is one very interesting example], fluid flow simulations)

3. Electronics in general that are constantly being pushed to the next generation in:
    a. Packaging to get more performance in smaller footprints
    b. Structural integrity
    c. Supplier selection for cost reduction

4. Wireless Service Providers & OEMs
    a. Analysis of waveform coding vs. environment (multi-path scattering from buildings, etc.)
    b. Efficient bandwidth (BW) utilization schemes (only so much BW allowed by the FAA so maximizing the number of users is of interest to the providers like Verizon)
    c. Antenna designs for robust coverage and gain

d. Denser electronics; higher performance; more functionality
5. Environmental
   a. Increased climate warming trends
   b. Waste management
   c. Alternate energy (please quiet the windmills!)
   d. Earth's temperature mapping from space
   e. Air column air quality, temperature, humidity, etc. evaluations from space
6. Automotive Industry
   a. Cars (software, Robotics, interior quieting, economy)
   b. Large Equip (Structural integrity, Load Estimation at weigh stations, automatic tire pressure adjustments, gear shifting designs with feedback loops, drive shaft torque)
7. Medical
   a. Software development
      i. Data base management
         1. information sharing among different medical groups
         2. linking between groups nationally and globally
      ii. Real-time testing/imaging of patients
      iii. User friendly interfaces for doctors.
   b. Acoustic probes
      i. Small parts (pancreas, etc.)
      ii. Cardiac
   c. Wireless products
      i. Writing notes in real-time
      ii. Signatures of medical personnel for authorizations
      iii. Sharing info in real time
      iv. Feedback to data management systems
   d. Electronics (Robotics, High V-caps for MRI, etc.)
8. Materials
   a. Nanotechnology
      i. surface physics,

   ii. molecular biology,
   iii. medicine
   iv. micro-fabrication
 b. Semiconductors
   i. Higher performing ICs
   ii. Active devices (amplifiers, mixers, etc.)
   iii. Diodes
 c. Ferroelectrics
   i. Mini-tunable capacitors
    1. heat/fire sensors
    2. fuel injectors (automotive)
    3. vibration detection
 d. Biophysics
   i. Molecular interactions (RNA, DNA)
   ii. Neural circuit (brain, organs)
   iii. Theoretical/experimental studies on
    1. ecosystems,
    2. tissues,
    3. populations
 e. Optical
   i. Tweezers (manipulating molecules)
   ii. Spectroscopy
   iii. Lasers
   iv. Interferometers
   v. Linear and non-linear dynamics
   vi. Uses for commercial, medicine and manufacturing disciplines

So the above short list provides examples of work that is currently growing and will continue into the future that you might want to consider. It can be aligned with the companies below, so here are some very exciting employers to consider exploring for work whether your interest is in the DoD, Wireless, Medical or another technical profession. The list is not inclusive by any means but is derived from my personal experiences which were very positive with each. There were also many smaller companies I dealt with over my career that conducted very interesting work as well, but due to space limitations I chose not to include them. It is no reflection on any of them as they were great to work with so my apologies to any of them who may read this book.

1. DoD funded activities (includes land/air/sea/space applications in different locations):
    a. Lockheed Martin
    b. Raytheon
        i. Also includes commercial airport development
    c. Northrop Grumman
    d. BAE
    e. Boeing
        i. Includes a large commercial sector with aircraft including lightning arrestors
    f. Harris Corp.
    g. Government labs (ONR, DARPA, SPAWAR, NRL, NUWC, Sandia National Labs, Livermore, Los Alamos, etc.)
    h. Applied Physics Labs:
        i. Johns-Hopkins APL
        ii. Penn State ARL
        iii. University of TX ARL
        iv. University of WA APL
        v. University of HI, ARL
2. Some Space Agencies:
    a. NASA,
    b. TASI (Thales Alenia Space Industrial),
    c. Jet Propulsion Lab
    d. Goddard Space Center
    e. Ball Aerospace
3. Automotive in USA:
    a. GM,
    b. Ford,
    c. Chrysler (now FCA USA),
    d. Nissan,
    e. Mitsubishi

      f.   Google driverless cars

      g.   Tesla Motors electric cars

4. Medical:
    a. Agilent
    b. Philips
    c. GEMS (GE Medical Systems)
    d. Becton, Dickson and Company
5. Wireless OEMs:
    a. Alcatel-Lucent,
    b. Ericsson,
    c. Qualcomm,
    d. Nortel
    e. Novatel
    f. Sierra Wireless
6. Wireless providers:
    a. Verizon,
    b. AT&T
    c. Sprint
    d. T-Mobile
7. Computer Hardware/Software
    a. SanDisk
    b. Dell
    c. Lenovo
    d. Microsoft
    e. Apple
8. Others
    a. Nanolink
    b. 3M
    c. Imagine Optic
    d. Master Bond Inc.

Some additional areas besides the ones discussed above, that you'll find interesting which utilize high tech skills are

- Journalism
- Sales and Marketing
- Law
- Finance

- Fashion
- Libraries
- Computer Science
- Astronomy
- Biology
- Geology
- Physical Chemistry vs. Chemical Physics

All of these areas mentioned above offer exciting opportunities and have tremendous potential for career enhancement. Some of them are niche positions but very often grow into rather large scale requirements. One example of this growth from the time I began working has been in the field of lightning protection for aircraft. It was a very niche area of research and product development but today is in demand and growing. (And as a multi-million mile flier and having been on aircraft several times when it was struck by lightning, I am glad to see this expertise develop!) It can be even frustrating (a good problem to have) if you find yourself struggling between offers in order to make a choice.

Figure 2 Hmmm What To Do?

So if the above discussion leaves you scratching your head as you wonder how you might really use your skills in private industry, remember "don't worry; be happy". Let's look at some examples in more detail that should help provide a clearer picture.

3.1   In-Depth Look at Some Career Opportunities.

There are a number of examples of different types of employment as indicated above. But to get a better understanding of how one can port their academic skills and interests into some of these areas, let's look at some examples in more detail. So let's consider a few examples of how one might use their skill set to support the needs of industry. The examples discussed below are not in any particular order.

<u>Inventory Management</u>.
This is a very important function in today's commercial as well as military operations environment. It often falls under the management of Manufacturing in the Org-Chart discussed below but in some of the larger firms could be a separate entity in the corporation. Imagine getting millions of components – some small some very large. How do you store them so that you can
1. retrieve them on a moment's notice,
2. find out how many you have at any one time,
3. know the lead time so you can replenish stock timely,
4. what is the planned usage so you don't have too many sitting around (they could go out of warranty before used, it is costly to have them sitting on a shelf, could corrode, etc.)
5. track costs so you know if an element from a particular supplier jumps drastically in one year vs. another; or better to purchase an element at a particular time of year
6. list the approved suppliers from Quality Control (more about QC below)
7. compare quotes for various volumes of expected uses so one can
    a. negotiate price with each supplier
    b. compare delivery between suppliers vs. your company's need
    c. compare warranty details
    d. identify any past or present quality issues
8. maintain a list of materials that can or can not be used in the product you intend to purchase (some could be carcinogenic, some can't be used in space applications, etc.)

And the list goes on but you get the idea. This position can be very interesting and there are some who enjoy developing the software tools in order for inventory managers to use. Others enjoy using the tools and doing the statistical analyses that support their organization as well as being in a highly active, large facility so that diversity of office and floor responsibilities keep the job excitement high.

But maybe you are thinking that is all useful information but how do you actually get the product? Some of these warehouses can be over one million square feet in size. In these large facilities such as at Amazon, Walmart, wireless product manufacturers, military installations, defense OEM contractors, shipping containers at an ocean dockside, etc. robotics are typically the answer in today's world. Here again you might find it more interesting to actually program the robot so that it knows all the detailed positions within the warehouse that each item is located. These items could be located on a shelf near the floor or several stories up. Some of the robotics are small some are incredibly large. Or, maybe you prefer to develop the robot and would work at a company that focuses on building this product. If mechanical, then the design of the actual mechanics of how it moves, etc. would be important. If software oriented, developing a tool set for the users of your product so it can easily adapt to their specific situation might be of interest. In either case it requires that you know a little about how it will be used at the various customer sites.

Semiconductor Processing.

This is a large sector in today's private industry and is used by both commercial and defense contractors. Its primary focus is to build micro-circuit based products. Some of these products can be a single, complex function chipset up to others which can be very large sixteen inch rack mounted products that might contain thousands of small chip elements. But there is a great deal of commonality in many of the processing steps. As such let's take a look at some of these overlapped areas and see what might be needed for a skill set to work in this arena. The semiconductor world turns out such products as HDTV, hand-held wireless devices, computers, missiles, advanced radar products, etc. There are a wide range of products.

As the demand for higher performance products and lower cost increased from both the general consumer as well as from the US Government on defense related products, there was increased pressure to accomplish both. Prior to the 1980s if products failed "out in the field", namely once they were delivered and in use, there was no formal analytical means available to access in order to use as a guide to trouble shoot, find the root cause and fix for future products. The method used was a combination of some crude data analysis taken during manufacturing along with intuitive concepts and assumptions. This clearly did not help to achieve either goal of providing higher performance nor did it allow for cost reduction. So, around 1986 Bill Smith, a Motorola engineer, developed what is termed today as 6-Sigma (you can find many references on this). Jack Welch later instituted this concept within General Electric around 1995 and after that bold move it was seen as not just a "Motorola concept" so many other companies adopted this method. Today there are a number of variations on this methodology.

There were basically three principles that were developed and implemented in the 6-Sigma doctrine which are:
1. There must be a continuous effort made to achieve stable and predictable processes which are vital to the company's success (e.g. reducing the variations which can occur in unit to unit during manufacturing).
2. Both the business (sales and marketing basically) and manufacturing elements of a corporation must be measurable, analyzed, controlled and improved through the processes implemented.
3. In order to improve the quality of output goods by a corporation requires the entire organization from top-level management down be committed to the effort.

As a result of implementing these principles, the 6-Sigma doctrine was implemented in order to improve the quality of the company's output by identifying and removing causes of defects as it attempted to minimize the statistical variability in manufacturing and business practices. A set of quality management methods were instituted which used a set of statistical methods. Today it has been used to reduce pollution, cut costs, improve throughput time on manufacturing (viz. faster delivery of the end product or process), fewer field failures, increase customer satisfaction and finally to increase profits.

The goal was to get to a 6-Sigma level where the product would be 99.99966% free of defects. It has never been achievable but continues to be referenced in today's world since the goal is to try to achieve this level. So this is basically the mindset in today's business world and keeping this in mind let's consider some aspects of design and manufacturing that has some very interesting technical aspects worth noting in more detail.

Engineering.
Once a job comes in-house through what we call a Purchase Order (PO) it goes to engineering for design. Here the engineering group could be a combination of mechanical design, electrical and system performance analysis. The system guys typically do an analysis of the environment and layout the method for the designers to follow to achieve the required performance of the product being purchased by the customer. This can be quite an involved process and to appreciate this complexity consider the following simple mobile phone handset.

Suppose you were in a field talking to a friend on your mobile phone. Most likely the signal from the base station to your handset would be a direct path and the clarity of your signal would be a function of your Signal-to-Noise ratio (SNR). Namely, if the weather was bad and you were outside in a heavy rain, this would absorb the microwave signal to your handset (typically near 2GHz). However, suppose you were on a sidewalk in the city with many high rise buildings around you. Now the signal from the base station will arrive at your handset from multiple paths. The signal could arrive at your handset via a single bounce (and this could be a single scattering or reflection from different buildings), multiple reflections from multiple buildings, etc. So the actual signal arriving at your phone in this scenario would be many time-delayed replicas (namely the signal from the base station could bounce off many buildings before arriving at your handset). With these time-delayed replicas all arriving at your handset, how do you process them so you don't hear a garbled voice message? Same is true when considering the reception of a radar signal. Or in the medical field, if the product is an ultrasonic probe which sends acoustic waves into the body, there could be multiple reflections from various internal human organs that arrive back at the receiver head and could be a confusing display of signals for the physicians to study and evaluate. To improve this capability and provide a better tool for the medial professionals, the system engineers continually study the various reflections that could occur, design various signals for more robust processing as well as display architectures. So whatever the field of interest one can appreciate that this aspect of engineering is not an easy task but absolutely a necessary one.

Once the system engineers (or scientists…typically in industry everyone assumes the title of "engineer" in some capacity just for brevity) have completed their modelling, the design team begins to work on the product. Usually a customer will define some bounds on the performance that is acceptable and could be defined as either a "not to exceed" bound or "must operate within these parameters" type bound, etc. These are perhaps the two most popular one will often see.

The design team knows the parametric bounds of each step in the manufacturing process within their company. These statistical errors are recorded and constantly analyzed within by manufacturing and an updated data base is maintained. This statistical data base is what the designers will go to as they progress.

Typically, the designer would begin the design with the "ideal" case to confirm that the product, if produced and used under ideal conditions (of course this is not realistic but represents the use of the product in an ideal environment along with no errors in manufacturing) can meet the desired performance required by the customer. If not, then an iteration process occurs in order to see if a design under ideal conditions could converge to the customer performance parameters. If not, then Sales and Engineering usually discuss trade-off parameters with the customer in order to be able to build a product (however, this negotiation usually happens during the early phases of quotation to the customer).

Once an ideal design is confirmed, the engineer then redoes the analysis adding in the statistical boundaries for each step to get the bounds on the worse case performance. Sometimes the tolerance bound for one step in the manufacturing process has to be controlled tightly in order that the product meet the customer requirements when completed. This would entail discussion with manufacturing to alert and get agreement that the tight control needed would be acceptable and achievable.

When the design is complete there is usually a "hand-off" meeting where manufacturing, Quality Control, sales, etc. meet and any concerns or cautions are discussed. Once completed and the design is accepted manufacturing takes on the role of building the product based on the design.

> **Side Note.** Even though the discussion is focused here on the semiconductor industry these parametric bounded concepts occur in almost every industry. For example, consider the fashion industry. When cutting fabric in order for the final product to meet the standards of a size "34 inseam" or a generic size small labelled "S" the fabric delivered to the seamstress must be within certain parameters. Even the seamstress has bounds on the equipment used in sewing – each fabric has slight differences when completed and is not

unexpected. So the equipment used for cutting, sewing, etc. all must be able to meet the bounds required by the industry. It is an important and interesting aspect of engineering and manufacturing.

Circuit Board Design.

So-called circuit boards can be the old version of Flame Retardant (hence the FR designation you will see on many of these items) glass epoxy boards to ceramics and other non-conductive elements that are being studied and used in special applications. They can be single layer, double layer (circuits on top and bottom of the board) or multi-layer boards (can be dozens of layers). The design engineer knows the bounds of the actual build of the circuit board. If it is to be built on the glass epoxy FR boards, typically a Mylar mask is made which overlays on the copper clad board which is then submerged in an acid bath. The length of time it remains in the acid depends on the thickness of the copper and the line width of the printed circuit. The statistical errors that are typical in this process is the over/under etch (namely the lines on the printed circuit board (PCB) could be "thinner" or "wider" than ideal). The difference these errors make in the total product performance is accounted for in the design.

Looking at this in some more detail one notices that there could be additional resistive losses due to thinner lines, or more inductive/capacitive coupling loss if larger. There will be pads where by surface mount components might be soldered or wirebonds attached. These pads will couple capacitively to those in layers below and above it (due to the finite permittivity of the board material) which will have an effect on the overall performance. These need to be calculated and the manufacturing engineers need to monitor very closely when making the board. If they are unable to meet the statistical bounds allowed for from the design it will impact yield and hence cost and could be an impact on profit if not accounted for correctly during the design phase.

Another popular means of making printed circuits is direct write using laser. Again, the statistical controls of lines, pads, etc. must be maintained and known. Additionally, product could be lithographically printed.

> **Side Note.** Most all processes in high volume manufacturing today are with automated equipment designed for special purposes such as wafer trackers, pick-and-place equipment (take chip components from tape and reel and place them at the correct position on the printed circuit), soldering (using wave soldering, for example), etc. Some of the high performance, high powered products for defense are still made with a mix of hands-on labor as well as automatic processes.

Wafer Processing.
This process is carried out in what are called clean rooms. Initially these rooms were designed so that air was forced through HEPA (Highly Efficient Particulate Arrestant) filters and exited through the floors into ducts underneath the floor where the air was vented out into an air handling system. Here the air was conditioned (removed humidity and cooled the air) before it was returned back into the ceiling ducts to be re-circulated once again. The volume of air flow as quite high and as the floor was a thin material sitting on the duct work it often "floated".

> **Side Note.** As a theoretician designing product once out of school, I wanted experience in the labs within the company to understand in each case the challenges every step in the manufacturing process required. It gave me a much better appreciation of the talent for each manufacturing engineer and challenges faced. I started my "tour" in the photolithographic section and was not aware of all the issues in a clean room. First it was pressurized (and still is) so this had a slight effect on me as I began to work. But then, and all the personnel in the room waited in quiet anticipation of this happening – a

sadistic joke I might add but now find it just as funny looking back – but I began to get sea sick. I struggled for a while and then could take it no longer and made a hasty exit as they all laughed and clapped as I had exceeded the time limit they thought it would take before I left. As I was the first in the company to work my way through all the steps necessary in manufacturing and testing a product I made a lot of friends and this came back to help me many times. But I learned that I needed to get my sea legs when working on a floating floor in the clean room which was standard in the 1980s.

Today, HEPA filters are still used in the overall room but the actual processes are performed in localized areas which are highly filtered smaller areas so easier to control than a large room and less expensive for sure. So each work station if you will is tightly maintained for humidity and temperature as well as air cleanliness. This air quality is given by a HEPA rating which is designed as a minimum to remove 99.97% of all 0.3 μm particulate and keep an air flow out of the filter at no less than 0.044psi (300pascals). As the percent of particulates increases so does the HEPA filter rating. The earlier classification of clean rooms was done as a number rating such as Class 100,000 which was the lowest rating. As the manufacture of these rooms became more sophisticated (and there are dedicated manufacturers that only build these rooms) they became more standardized and was adopted into an International Organization for Standardization (ISO…a European format). This new standardization and the earlier Class XX (noted as FED STD 209E) format are listed and correlated in the table below since if you work in the semiconductor type environment you no doubt will come across this. It is referred to as ISO 14644-1 Clean Room Standards.

Class	maximum particle size/m³						FED STD 209E
	≥0.1 µm	≥0.2 µm	≥0.3 µm	≥0.5 µm	≥1 µm	≥5 µm	
ISO 1	10	2.37	1.02	0.35	0.083	0.0029	
ISO 2	100	23.7	10.2	3.5	0.83	0.029	
ISO 3	1,000	237	102	35	8.3	0.29	Class 1
ISO 4	10,000	2,370	1,020	352	83	2.9	Class 10
ISO 5	100,000	23,700	10,200	3,520	832	29	Class 100

More data can be found in the literature but what is interesting to recognize is that room air or ISO 9 (not shown above) is basically unfiltered air. In the medical industry, the use of clean rooms is very important and parallels the above. For many applications there is the added step of using a UV light source that kills live bacteria and viruses trapped in the filters prior to entry into the room being conditioned. This added control of airborne bacteria and viruses is an attempt at the prevention of or spread of infection (imagine having a medicine prepared with unwanted bacteria present…hope that doesn't end up in any of us but we all hear of these types of cases on the news every so often!).

Side Note. In the 1970s and 80s NASA needed higher performance electronics in order to carry out some of its missions. From this need a research effort was conducted in order to determine the best means possible to accomplish this. It was discovered that under the current conditions of the time, dust was a large contributing factor that limited capability. As such the "clean room" was developed. Today this has spilled over into the medical industry and is used to construct operating rooms.

So let's consider a process flow for a wafer as the manufacturing process begins. Please keep in mind this is a rather specific example as there are many variations in the industry depending on the end result desired. All raw materials start in Inventory Control which we discussed above. The first step would be to retrieve the wafers from inventory control and these would be circular wafers (could be quartz, silicon, etc.) which might be as small as 3 inches in diameter up to 12 inches with varying thicknesses. The process steps could include doping or ion implantation, deposition of one or more layers, etching or photolithography. This would be followed by dicing and packaging at the end before going into a chamber for sealing. The end result from wafer processing is to make a microcircuit, chip or chip set (multifunction capability vs. a chip which might be designed for just a single function). Figure 3 shows a set of wafers stacked in a carrier ready to be mounted in a wafer processing machine.

Figure 3. Wafer Carrier

The intent will be to get a printed circuit onto the wafer and this circuit would be a tiny percent of the overall wafer size. So we would reproduce this image many times over the surface of the wafer. This process could be done in a single step using a multi-image mask or a single image mask and then stepped across the wafer. In some of the more sophisticated processes, a portion of the overall circuit desired would be stepped, then another and another, etc. This would build up the microcircuit images on the wafer through many steps. This can get quite complicated as you might expect so for us to get an appreciation of the wafer processing we'll concentrate on a single image mask.

The output from the design engineers would be a mask which is a pattern that is burned onto a quartz plate typically. This pattern would consist of what we will call a single chip and again this is quite small compared to the size of the overall wafer being used. So we can duplicate this image many times (sometimes thousands or more). Since our plan is to duplicate this image multiple times on a wafer, we will need to cut each die and mount in its own package. Since this cutting or dicing as it is often called, can only proceed in straight lines the die themselves have to be in a rectangular patter. To get an idea of the number, a rule of thumb is

(1)   Die per Wafer $= \pi d^2 / (4 A)$

Where "d" is the diameter of the circular wafer and A is the area of the die. So depending on the size of the circular wafer and the extremely small size of die used today in our microcircuits, the die per wafer can be very large! The die are typically separated by a small bias distance which is the width or kerf of the dicing blade.

So we have a mask designed by engineering and a wafer so the first step is to decide which of the two photolithographic processes we want to use. These are termed "Wet Etch" and "Lift-Off" and illustrated in the figure below. They are characteristically different in that for the Wet Etch one would metalize the wafer first and then add photoresist whereas for Lift Off it is opposite in that photoresist is added first followed by metallization.

Photolithographic Fabrication

Etching Techniques

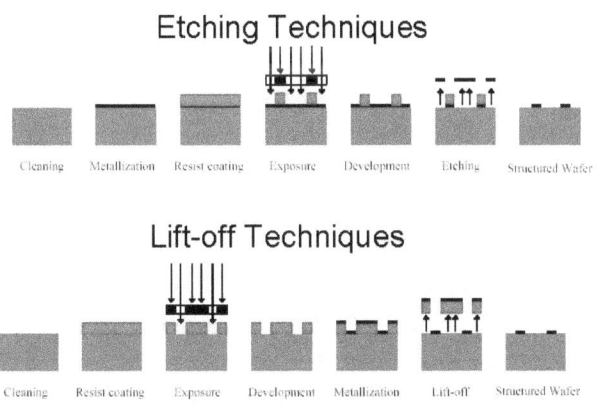

| Cleaning | Metallization | Resist coating | Exposure | Development | Etching | Structured Wafer |

Lift-off Techniques

| Cleaning | Resist coating | Exposure | Development | Metallization | Lift-off | Structured Wafer |

Figure 4. Photolithographic Fabrication

Discussing the details of the processes or the differences is not the scope of this book but is noted to introduce you to an important process that offers challenging technical issues to solve and is in a very large sector of employment, the semiconductor industry. For now, capturing the concepts and getting a high level understanding regarding the various skills needed should be appreciated so you can evaluate if it meets your interests and skill set.

As photoresist is light sensitive you can identify the rooms where it is performed by yellowish or gold tinted windows in observation rooms. There are many technical challenges that make this phase interesting in that it is temperature, humidity and light sensitive. Trying to uniformly coat a wafer is a real challenge as is the process of either wet-etch or lift-off. Non-uniform coatings affect yield which affect cost which affects profit. Variations in wafer thickness vs. the required thickness for the specific microcircuit to be manufactured are carefully tabulated and made available for the designers. If you enjoy both analytical and hands-on work that entails strong chemistry this is an exciting arena.

Sealing.

Typically after a wafer is photolithographically processed it is stuck onto a tape and is then diced into many pieces which are the die. These die are transferred over and glued into packages (thermal stresses, second order effects, etc. all become extremely important and can be exaggerated at this point if not careful). The package is generally sealed using a thermal seal (non-hermetic), solder or welded (which could be resistive, cold weld, etc.).

Assembly.

From here the product could be integrated into higher order assemblies and onto larger circuit boards. This could be done manually in some cases or through automated equipment.

Test.

Throughout the flow of work steps in the manufacturing process, there are tests done to make sure the partial assembly performs the way it should before it is passed on to the next phase. This is especially important at the initial wafer processing stage since once you have a multi-image wafer ready for dice and assembly, one probes each die with electrostatic probes typically to test if the die are functional or not. Those that are not become discarded and those passing this test are assembled.

If by now your head is filling up with information and you feel it pounding remember the story about the patient who visited the Doctor and complained of a problem. The Doctor noted "Don't worry, we'll have you back playing piano in no time." And the patient smiled brightly and said "That's great, since I always wanted to play piano and never did before". So let's move on and see where some of the job growth is occurring in today's market.

Medical Field.
There are many applications that can be considered in this industry. As you may know, the medical field is growing very rapidly and has many exciting opportunities for technical skills. It is also very rewarding to know your work is directly helping to save lives. In addition, one gets to work closely with the medical professionals who actually will use the product and hear from them first hand what they need, their challenges, discuss trade-offs of design vs. their desire, etc. Let's look at two of them:

Magnetic Resonance Imaging.
If you have ever heard the "clang" of an MRI system what you hear is the discharge of high voltage capacitors that are in rings around the table the patient is lying on. This discharge generates an electromagnetic field that penetrates the human body and from the return field strength can display a very detailed structure of the body under investigation. Since the magnetic field strength is an inverse function of distance-squared and can range as high as 3T. Due to this rapid decrease in field the tube the patient lies in is rather small so the E&M field can penetrate the human tissue (mostly water which absorbs this radiation).

There continues to be work on high voltage capacitors and there assemblies to reduce the "clang", improve the strength of the field generated so the tube can be enlarged (some patients are very claustrophobic). In addition the entire system design, integration and assembly is highly dependent on technically skilled people.

Ultrasonic.
Another very rapidly growing field of medicine as it offers in office evaluation tools. These typically require the

design of a ceramic acoustic head that is diced into rectangular pieces and reside on a surface at the end of the probe. These probes can be used in non-invasive procedures such as cardiac imaging, small parts imaging (spleen, gall bladder, etc.) or for invasive procedures such as prostrate exams and biopsy samples.

The ceramic material is selected and diced based on the frequency of operation which itself is a function of the intended use and distance of operation (namely how far away is the expected target typically?) expected. What is interesting to note is that the ceramic elements are arranged in an array that allows for beam forming so as to minimize side lobes (which allows for spurious signals to enter the system and clutter the image). These elements have mutual capacitive coupling and this needs to be modeled in order to maximize performance. The assembly is very intense as there are strict regulations especially if the probe is to be used invasively.

When looking at the impedances of the unit, the ceramic head is of high impedance and the human tissue is at a lower impedance. This means that without a smooth transition, there could be a large reflection of signal back from the human. To counter this, a gel is used that helps to act as an impedance transition medium so that the maximum amount of energy from the probe will enter the patient.

Again, the mechanical design of the probe has to be visually appealing, easy to handle and functional. This includes the plastic probe head as well as cable since if the cable was too thick (generally to beam form a wire from each ceramic element to the receiver would be needed and for narrow beamformed images this could be quite a few wires). One can think of the beam forming as a series of Fourier elements…the more elements one has the steeper the main beam. This means if you Fourier transform a uniform strength set of elements (basically looks like a "square" in time as all the voltage strengths of each element are the same) you would get a Sinc function. The main lobe of the Sinc would get steeper and the width of the Sinc would shrink to become narrower as the number of elements (and wires!) increases as shown in Figure 5. Very good for imaging....very difficult to handle practically because of the size of the cable connecting the probe to the equipment. As a result it never made it into industry

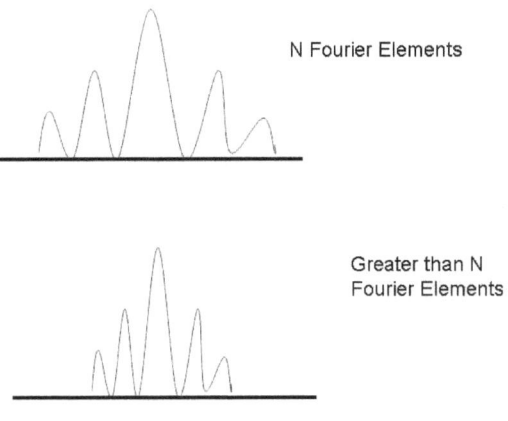

Figure 5. Sinc Function

3.2   Overview of More Career Possibilities

In Section 3.1 many detailed examples were provided in order to help identify where some of your skills could be applied and how they could be used in order to solve problems that exist in the marketplace today and is of interest to various companies and markets. Let's look at a few more opportunities but at a higher level since once you see what the needs and functions are it will be easy to connect your capabilities and interests to them should you chose to pursue. The list being referred to is duplicated here from above for convenience.

- Journalism and Communication
- Nanotechnology
- Sales
- Computer Science
- Law
- Finance
- Fashion
- Libraries
- Astronomy
- Biology
- Environmental
- Geology
- Physical Chemistry vs. Chemical Physics

Journalism and Communication

In today's fast paced world, we all tend to think in sound bytes. If the first few sentences of a written document or talk are not of interest we generally move on to something else. The ability of being able to digest technical information, reassemble and transmit to get the key elements out quickly and in an interesting format to capture attention is a very important capability. Not only does this require knowledge about the product but also about the intended audience. If the audience is comprised of mostly business people then the presentation whether written or verbal

45

must be at a high level but in such a manner they can capture the importance and understand the market better. If, however, the audience is more of a technical background then of course they want to understand more details about the nature of the product or products being presented. Having the technical and communication skills necessary to assess the venue of interest and structure an appropriate presentation is an important attribute in today's world.

This category could be filled by trade magazine journalists, reporters, technical writers, video presentations, and so forth. Being able to understand the technical jargon and use it appropriately is important and is found in such work as:

- technical self-help web pages,
- illustrations (there are some comical ones that are very effective out in the industry),
- FAQs with appropriate answers,
- comparing technologies,
- corporate reports,
- copywriting (e.g. ads),
- corporate roadmaps,
- free lance journalists (often to help foreign scientists write books to ensure that correct grammatical structure is insured but in a manner so as not to diminish the technical aspects to be presented),
- proposal writing

If you have a desire to interact with people, there is a need for technical specialists at science museums, libraries, planetariums and other public forums.

Nanotechnology

As many already know, nanotechnology is a branch of science which manipulates matter at the atomic or molecular level. The dimensions at which this occurs is between 1 and 100 nm (nanometers = 10^{-9}m namely one-billionth of a meter). The Hydrogen atom which we discussed at length is about ¼ nm. So how can one get a more intuitive "feel" of what a nanometer is? Consider a strand of hair. Now carefully slice down parallel to the length axis of the hair and keep cutting the remaining larger piece in this manner into about 100 equal size pieces so that you have about 100 long skinny pieces of this hair strand. Each portion would be about 1nm in thickness. It is then no surprise to recognize that Quantum Physics plays an essential role. From our ability to work at such small scales arise several very important applications as discussed in the following sections.

Nanotechnology in Medicine

Let's look at a few developing areas in the field of Nanoscience for medical applications. These areas will undoubtedly be well funded in the future and offer the opportunity to conduct some very interesting research as well as product development that will help millions of people.

Medical Drug Transport

One application of nanotechnology under development in medicine today involves the use of nanoparticles to deliver drugs, heat and other substances to specific types of cells, such as cancer cells. These nanoparticles are developed in such a way that they have the ability to identify unhealthy cells and attach to them allowing for the direct treatment of these cells. In this way, risk of damage to healthy tissue and cells in our bodies is minimized and could be used as a preventative means, namely they could be injected into our body much as we do now for flu shots where these engineered nanoparticles travel our bloodstreams looking for the unhealthy cells they were programmed to detect.

Researchers are also exploring the use of nanoparticles to detect sheer forces in our blood stream such as those that occur in an artery that could be blocked by a blood clot. Once these sheer forces are detected drugs would be released to dissolve the clot.

Nanoparticles that have a gelatin like structure are being used now in lab settings that have shown significant progress to repair damaged brain tissue. This method appears to be more efficient than current methods that are used in today's medical treatments. The hope is this will be a more efficient means of dealing with brain trauma in the near future.

A matrix of nanoparticles is being explored which will act as a sponge that will be used for insulin release. In this case, the sponge will first be impregnated with an enzyme before being injected into the body's blood stream. When the glucose level exceeds a set amount for the patient the nanomatrix releases positively charged ions that bind to fibers in the matrix. Since these are like charges, they repel and this creates a hole in the matrix through which insulin stored in the matrix is released into the blood stream.

Therapeutic Applications

Today an effort is in progress using the above mentioned concept of nanosponges. These sponges are coated with a red blood cell membrane which allows them to travel through the blood stream where they are "programmed" to absorb toxins and once found, remove the toxins from the body.

Sound waves have long been used in medicine such as the very familiar ultrasound imaging we often hear about. But today, carbon based nanotubes are being studied to be used for noninvasive surgery. Nanotubes today are grown in a lab and looking at them under an SEM (Scanning Electron Microscope) they actually look like a long hollow tube, in case you have never seen an image of one. A medical professional with the aid of a laser would shine light on the nanotubes and these particular nanotubes would then convert the laser light into highly focused sound waves. The intent is to demolish tumors and other diseased tissue using sound waves without harming healthy tissue.

Heat therapy is being studied in the attempt to combat breast cancer tumors. Each specific type of breast cancer has a very well known protein associated with it. Antibodies would be attached to nanotubes and the antibodies would be selected that are strongly attracted to the particular protein of interest. These nanotubes would then collect at the site of the tumor allowing light from an infrared laser to be absorbed and incinerate the tumor tissue using highly localized heat.

Nanotechnology in Electronics

There is a wide range of research being conducted in this area but here are some key developments underway:

- Integrated circuits are being engineered using silicon nanophotonics. This is an optical process which would allow for higher speed data transmission between electrical assemblies. It has a vast reaching range of applications in use for medicine, computers, automotive, wireless products, etc.
- Switches using nanomagnets are in development to replace some transistor applications in various electrical systems. This would mean longer lasting batteries as they would require significantly less power. We all would appreciate this as we use our laptops and wireless devices.
- Copper nanopaticles are being developed as a lead-free solder that could be used in space missions offering higher reliability in extreme environments.
- The use of carbon nanotubes is being applied to direct a bean of electrons which will illuminate the pixels in a very thin (less than 1mm thick) light weight display screen.
- Very dense low powered memory devices are underdevelopment using nanotechnology. The intent is to use an iron-nickel coating along a nanowire and a current would magnetize sections of the wire which would travel down being read by a stationary reader.

- A nanoglue is being studied that would couple a computer to a heat sink helping to keep the device cooler. This would save on the batter as well as the internal circuitry by running cooler and nanotechnology would even allow it to run faster – a process which usually generates more heat.

Nanotechnology in Space

Advances in nanomaterials could mean that space craft would be made of material that is much lighter allowing for lower fuel consumption on reaching space. These materials along with other products from nanoengineering such as nanofabrics for space suits or nanosensors and nanorobots to improve space ship performance, could make the reliability for space exploration very attractive. Carbon nanotubes might be used on the fuselage for the spacecraft so that it would be stronger and lighter in weight. Space suits would have an outer membrane of nanorobots so that any holes detected would be repaired. Thrusters are large and expensive today but through the use of various components from the world of nanoengineering these thrusters most likely will be replaced to reduce size. And nanosensors would be deployed in a strategic fashion to detect the presence of any harmful chemical important for the health of the astronauts.

Nanotechnology in Sports

Good news sports fans…nanoscience is also being applied to Tennis and Golf, in case you are interested in these sports. The applications here are:
- Nanoparticles are used to correct any imperfections in the shaft of clubs and rackets in an attempt to make the product more uniform and would in theory improve one's swing (any help is welcomed!).
- The strength of rackets and clubs are being increased by adding carbon nanotubes in the metal frames. This presumably will improve one's control over hitting the ball.

- Nanotubes are also being integrated into tennis ball materials so they will lose less air and provide a more consistent bounce as well as last longer.

Sales

This area offers a broad range of opportunities that can provide job satisfaction. So let's look at how this might happen more closely in the high tech industry whether it is commercial or defense related.

Today it is important for companies to get their product information out into the market place in a timely, cost effective and efficient manner. It must be easily understood and concise as the engineers at customer sites do not have much time to spend trying to figure out what it is that your company is offering, how it can be used and is it applicable to their problem for their company's product. So this effort closely parallels the Computer Science category to follow and the Communication area noted above.

One position in sales might be in-house sales where you answer questions from customers who might call in for product information, which is often the case where catalog items are sold such as with large distribution centers. In addition to calls one might be answering questions via social media, email or video conferencing (Web conferencing, Face Time, Skype are examples). It is called in-house sales since you basically work out of an office and this could be your home office in fact. Telecommuting has many advantages for companies today and often you might find that you might be allowed to work from home say one day per week or more depending on the company's location.

In other companies that offer a design-to-specification capability, sales literature for a company is in the form of web pages today with less and less copy printed. (It is interesting to note that print copy is still very popular overseas and is sort of viewed as a gift when provided which can be just a one page ad on a new product up to a full catalog.) The person in this sales position must again have technical skills to understand the product, know some of the history of where it has been used before in market places, know the competition and your products' strengths (and weaknesses as these will be pointed out to you by customers and you need to know how to respond). In addition this position requires someone to:

- have an interest in travel,
- be capable of working independently,
- think on your feet,
- have a desire to meet people,
- interest to move from a detailed technical career to a more business focused path,
- be pleasant to deal with under all circumstances (exceptions are few on this since that person might leave and you don't want to have a bad reputation which will be passed onto the new person coming on-board.) You can be firm in your position but keep it pleasant.
- know prices and your competitor's prices,
- have the ability to listen to the customer's needs (don't try to offer square pegs to fit their need to fill round holes),
- be in a position to assess if the customer you visit is really considering your product and this means you assign a probability to this opportunity,
- if in the military sector know about ITAR (you will be briefed on this internally at your company so don't worry about this for now),
- demonstrate confidence and develop a strong relationship with the customer so they will follow your lead and want to work with you.

Even at trade shows the technical sales people at a booth within the show would hand out DVDs or thumb drives with company names printed on them and the e-format of their product catalog contained on the storage media. Usually an ad would have an introduction on the category of applications which the product could be used (especially true if you knew your market well) and then a breakdown of individual products along with data sheets. This allows customers to evaluate for their needs and often a file called the S-parameters (if the product sold is of an electronic nature) would be available for download allowing the customer to insert into one of many circuit design programs available and see the performance that might be achieved using your company's product.

Computer Science

Clearly if your major was in Computer Science then clearly this is your focus in obtaining a career. But if you are in one of the Sciences, Math or Engineering fields you most likely will find companies will be interested in hiring you to utilize your computer skills to help solve problems. Here are a few examples where you will find programming and hardware knowledge will be needed.

As product flows through manufacturing lines it eventually comes to the test lab where it undergoes either a "go or no-go" test (this is usually done for low cost, high volume product) up to full blown Acceptance Test Procedures which are very extensive and agreed upon between the customer and your company at the time of order. In any case testing procedures are automated and could be run over various temperatures, under different environmental conditions such as shock and vibration, subjected to different levels of input power, etc. These various tests are coded into the measurement apparatus and these various pieces of equipment are tied together via busses which are then driven by a master code in line with the requirements for the test procedure. For the low cost high volume product lines these tests don't often change. But for "design to specification" products as noted earlier these will change and will be a function of the product being built. In addition, there are many areas within test labs that are constantly being programmed for R&D, production, new equipment, improved analytical modeling, etc.

In the medical industry, imaging software is very powerful and requires working close with the medical professionals who use the product. Their need is to be able to see and diagnose a patient quickly through user friendly displays. This is a real challenge as more and more data are made available through faster chipsets being used. In fact how to manage all the data from a machine is a challenge in itself. User interfaces and displays are very challenging areas and growing.

As noted in the Sales category above the company's main focus for advertisement is its web page. There is a very large need today for skilled web designers at companies and how to link these sites so they appear at the top of the results displayed by a search engine.

Gaming industry is all about computing skills as faster video continues to be built allowing for more details on the screen making the virtual experience more "real". This also folds over to the film industry where special effects are very much in high demand. Speed is also bandwidth driven. Signal bandwidth however is limited by the FCC so superior coding schemes to pack as much information into the allowed bandwidth as possible is in high demand.

The government sector also needs computer related skills. Some examples are the need to program stop lights in cities and towns to maximize traffic flow or writing search engines for archived data bases. If in a defense environment, very specific software is needed and is classified.

The automotive industry is really driven (sorry, couldn't resist this) by automation. Not to mention cars themselves. Today each car will have typically over one hundred microprocessors all tied into the master computer on-board.

The trucking industry uses code to smoothly shift between gears so as to reduce wear on moving parts, improve fuel economy, manage the increased loads these vehicles can handle, etc. Also, at weigh stations trucks roll over scales and quickly stabilizing the readout of a scale as a truck moves across so it can be read to with the statistical error allotted is very important. This is all done with the integration of hardware and software.

Analytical modeling for many markets are conducted by consulting firms where either the engineer/scientist would develop an analytical model and then write the software or receive the model from another department and write the code. This could be a study of optimizing the management of ships coming into port, traffic flow, people management schemes at large events (e.g. at Disney World or SeaWorld in FL where optimizing the flow through the park and at each event is studied) or theatre management in a battle group scenario. There are many types of these companies for many different applications.

Law

A grad student in Physics where I went to school and with whom I remain good friends took a position upon graduation with a large firm in the US. As it turned out our overseas travel plans while on business crossed paths one summer and we decided to meet for dinner at a nice restaurant along the Cote d'Azur. While eating we talked about our work and compared to our expectations when we were Physics grads. He actually did not like his career and told me he was going to make a drastic career change and "re-invent" himself. That he did as he went back to school and got his Law degree specializing as a Patent attorney.

Side Note. A patent is considered to be a form of intellectual property. It is consists of a set of exclusive rights granted by the government to an inventor or assignee that is in effect for a specific time period (as of May 13, 2015 it is fifteen years from the date of issue) in exchange for detailed public disclosure of an invention of a product or process. The procedure for granting patents and the extent of the exclusive rights vary widely between countries depending on the international agreements in place. The exclusive right of the patentee sought is to prevent others from commercially manufacturing, using, selling, importing or distributing the invention without permission. Patents are typically granted for those inventions that are novel and useful so a careful review of all existing knowledge and careful wording of the invention during the application phase is crucial for receiving a patent award. There are basically two main parts of the patent which consist of a description and the claims. The description must contain sufficient information for someone who is skilled in that technology could in fact build the product or duplicate the process based on the information provided by the filer. The claims section must detail what is being described while meeting the requirements for filing. The wording here must be carefully chosen to offer the widest possible protection for the patentee.

There is still a shortage in the patent field industry as technology is expanding rapidly. Some of the duties include researching existing literature, build strategies to protect the intellectual property of the patentee and file new applications for patents. Without a law degree you would find that most of your work would be centered on researching existing literature. It would be at the frontier of technology and you often would be the first outside the company filing for a patent to read and understand what the intent of the invention or process is about, which can be very exciting. It requires that one work quickly to afford the best opportunity to file first as well as working as part of an elite team.

Positions in the patent field that require researching literature range from the BS to the PhD and of course pay is commensurate with the degree along with experience. There are entry level positions available to consider and interestingly many are located along either the West Coast or East Coast of the US due to the high volume of export and import. But the majority of positions are within law firms, large corporations, universities and some technical experts even contract out their services (but this takes a few years in order to develop contacts).

Finance

The analytical skills and ability to draw conclusions based on data analysis is an important aspect should you desire to enter into the world of finance. Actuarial science is an area that is of great importance not only to Wall Street but to corporations and universities alike. To become a fully licensed actuary requires passing an exam and there are many study courses available to accomplish this. It is a profession that is consistently ranked as one of the most desirable (Ref. Riley).

As an actuary you become a business professional (reporting into the Finance side of the business) dealing with the management of risk, its measurement and uncertainty. Mathematical skills (typically calculus, statistics and probability) are important in this application along with computer science, economics and of course finance. Quite often actuaries are employed by insurance companies to analyze and assess the probability for death, injury, illness, disability (long and short term) and loss of company property. This information on a measure of risks, mitigation of risks and the statistical uncertainty of the analytical model are considered when underwriting an insurance and employee benefit plan for a corporation.

Another area that actuaries are often used within companies is to analyze the level of resource investment while minimizing risk as a means to maximize profits. This requires looking out several years based on marketing plans in order to estimate the return-on-investment (ROI). As businesses consider moving into new markets this is an important consideration as it mostly likely be out of their developed core competence. Analyzing and interpreting this analytical model is an important aspect for upper management to make a decision. It will have a direct effect on the balance sheet.

Fashion

A close friend is an executive at Victoria's Secret and has had a fascinating lifestyle in the fashion industry. The tools used in marketing and data analysis are highly sophisticated analytical tools. The results of their constant studies are incredibly detailed and demonstrate how well they work when diligently applied as Victoria's Secret is an extremely profitable corporation. For example, they study which colors are more popular in various locations throughout the country; when do these colors sell best; what styles vs. locales are most popular; and to help bring the men into their stores by themselves to generate sales, they position specific articles of clothing at the front of the store and this changes as you get deeper into the store. In addition to the position of their clothing line there is careful consideration given so 'friendly' color choices and styles are displayed up front. Not only is the position of lingerie important but the sales associates working at the front have different sales skills compared to the ones at the back of the store can serve a different function. Today based on detailed market studies I understand the training for their sales people has significantly evolved into a program that is focused to turn out employees in what are termed now as "highly efficient sales associates". The analytical marketing performed to implement this detailed approach has produced remarkable results. But I digress as this is another story … so back to the use of technical skills as applied to the fashion industry. (Incidentally, if I got any of this wrong it is my misunderstanding and in no way reflects on information that was explained to me privately.)

One interesting side to note from the above discussion is the use of analytical tools/methods in high tech companies is very much an integral part of non-tech companies in general. Given competition, consumer demands and the desire to maximize profits, there is an emerging profession within the private sector which is often referred to as "Data Scientist". It is in fact an advanced version of the former "Quantitative Analyst". To fill this need in private industry, computer science departments are developing courses that specialize in two areas often referred to as "Big Data" and "Machine Learning". Here is a brief explanation of each.

> (1) Big data refers to all information collected in the format of some data collection protocol. In the example above with Victoria's Secret this might refer to the revenue of all of the stores national-wide, the popular colors and times of year sales peak, etc.
> (2) "Machine Learning" uses a statistical analysis to first evaluate the data and more importantly to forecast some future trends based on that. In our example that could be the most popular color vs. season for example. Therefore for the following year the stores would arrange the delivery of new products accordingly to maximize profits and hopefully increase market share in the industry. Currently, the data files of information collected are becoming more complicated as they are becoming multi-dimensional. Namely there are a number of parameters that need to be attached to the data collection process which in turn requires the analytic tools to become more advanced.

This need will become very much in demand in the coming years. There are many more fashion companies to be considered as this market is growing rapidly. If fashion peaks your interest you could find that you will have a long prosperous career. It is fast paced, rewarding with rapid results and lucrative.

Libraries

When I was finishing my Doctorate degree and writing a bibliography section for my dissertation I spent a great deal of time in our school's library. As it turned out, the university had just built a new library building on campus and as my luck would have it, they were in the process of moving from one to the other so it was necessary to conduct my literature search between two buildings. But some would say those were the good ole days...yeah sure. Anyway the system in place was the Dewey Decimal system which used index card files...and thousands of them. So you would flip through these cards, read a short abstract that somebody had hand typed on the card and if it sounded interesting you would then write down the reference number and go look for it in the library. It worked as long as everyone replaced the reference book or magazine back on the rack in the right place...otherwise, well you get the idea. So you probably are thinking as you read this that this system was just one step up from chiseling on slate tablets and perhaps it was. However, it worked and the librarians at the time were extremely efficient at managing this system and keeping most things very organized. Today, one would search electronically for references that might be in hard copy at the library or in an e-format. Many of the public library users are not computer literate so there is a need to help teach or at least help find material of interest for the person. Having someone who is well versed in the computer systems at libraries and can help teach or find reference material for users is a great asset today. Some websites will in fact charge to download magazines for example and if the article of interest is not in that magazine, well a few times making this mistake can get costly to users not to mention frustrating. So having technically savvy people at the library to help on this is a great value.

Astronomy

Even though NASA funding is down over the last few years here in the USA, there is a global effort that has actually grown space borne applications as well as research. This funding has been supplemented by the European Space Agency (ESA) funneled through various major design centers and Japan for the most part. There are other supporters but these are the largest two contributors along with NASA in the free world economy. Some of the research areas and applications can be fascinating but the research programs are the markets that astronomers would enter. In addition to Space agency funding are public organizations and private sources. Some of the duties that would be conducted by an astronomer are:

1. Analyze research data to determine its significance.
2. Teach astronomy at schools and universities.
3. Work at a planetarium or scientific museum.
4. Write proposals for funding of various astronomical studies.
5. Evaluate orbits and determine the size, shape, brightness and motion of different celestial bodies under study.
6. Collaborate with other astronomers to carry out the research and this usually entails an international consortium.
7. Support the development (or be the originator) based on observations.
8. Measure radio, infrared, gamma and x-ray emissions from extraterrestrial sources.
9. Present research findings at conferences, write papers for technical journals.
10. Discuss advances in astronomy at public forums and support related programs.
11. Study, develop and support the use of ground-based and space borne telescopes and scientific instrumentation.

12. Design, develop or revise current instrumentation and software for observation and analysis of astronomical data.
13. Support the development and implementation of flight plans and satellite communications.

Biology

If you are not a biologist then you most likely will find that you would fit into a category generically referenced as a Biophysicist or Biochemist. The primary responsibilities can be very similar and so we'll keep it general in our discussion that follows. Each job in any field will have specific details that would be impossible to cover in a book but by understanding the primary role of the basic job description you can determine if the category is right for you as you consider "What do you want to do when you grow up?" (I'm still pondering this question!). Many of the employers in this field are pharmaceutical companies, universities, state agencies and government labs (Lawrence Livermore for example).

The primary role of in this function would be to study, analyze and report on the physical and chemical properties and principles of living entities and biological processes. In this pursuit typical equipment used would be lasers, electron microscopes and other more common lab equipment. Biological samples are collected which are then scientifically tested and analysis is then performed on the samples. Often one might find the need to use chemical enzymes to synthesize recombinant DNA or study extremely volatile chemicals, viruses and bacteria (usually falls into the study of microbes). This latter study might be an estimate on which flu is expected to be prevalent so a vaccine can be developed or a bacterium that threatens world health (we've all heard about the anthrax issues).

Often computer modeling is performed to determine the 3-D structure of proteins and other molecules for various needs.

Professionals in these fields observe the effects of drugs, hormones and food on tissues and biological processes. This could also lead to genetic mutations in organisms such as cancer or other diseases. Important today is the study of Alzheimer's disease and the development of new drugs and medications.

One could get involved with agriculture through the study of plants and animals to understand the genetic traits carried through successive generations. This will help in breeding programs, developing the best feed, least harmful insecticides to use, etc. Developing genetically engineered crops modified to resist drought, disease, insects, etc. while increasing yield per acre.

And an area that is in the news over the last decade and will remain so for many years is the need to obtain alternative fuels which might be biofuels derived from plants. A continued effort is also underway to help protect the environment and clean up pollution (such as that resulting from oil spills and carbon based emissions from coal fired electrical generation plants).

Environmental

Employment in the environmental market is expected to grow by about 10% over the next ten years, which is very aggressive. It is opened to all technical disciplines including Physics, Chemistry, Biology, Math/Statistics and Engineering and originated from an outgrowth of studies in Natural History and Medicine. The responsibilities typically include environmental studies, modeling and test/evaluation on soil science, efficient means of recycling refuse, public health concerns, waste disposal and water and air pollution control. Those with an interest in the pure science aspects might find it interesting to pursue environmental science where there are studies conducted on the underlying principles which govern mass and energy transport in a soil-plant-atmosphere system. Mathematical models are developed that describe the behavior in terms of environmental parameters which might include radiant energy, temperature, concentration of water, nutrients and other compounds whether natural or manmade.

If you search for these opportunities you will find that there are companies, government agencies and universities studying environmentally related issues ranging from water and contamination transport in soils to the issues on the impact of global climate changes on our environment including the long and short term effect on our ecosystem. This also includes studies in alternate energy sources that might be in the form of physical, chemical or biological processes.

Atmospheric sciences can include studies in greenhouse gases, meteorology, atmospheric dispersion modeling of airborne contaminants (volcanoes, smoke stacks, etc.) and sound propagation as it is related to noise pollution (low frequency energy from windmills is one example where it has been found to affect nearby neighbors).

In the field of study regarding the concept of global warming models are continually being developed to simulate atmospheric circulation, infra-red radiation reflection and transmission which are all then compared to test results. Chemists will find an interest in studying atmospheric chemicals and their reactions to our ecosystem whereas biologists analyze plant and animal contributions to the carbon dioxide content of our atmosphere. Those interested in oceanography will study the effect of absorption and re-emission of gasses at the surface and its effect on environmental temperature. They also study the effect of the warming trend on the polar ice caps which play such an important role in our life on Earth.

Side Note. One important area that scientists have discovered about Earth's polar ice caps which points to the importance of life is its color and structure. Ice is a natural reflector that bounces the sun's heating rays back into space and helps to keep Earth cool. As the ice caps melt there is less reflective surface of course. The ocean and land masses, contrary to ice, absorb more of the heat from the sun than reflect it harmlessly back out into space. And of course this increased temperature induces a more rapid melting rate. This is the main reason you hear in the news today about the concern on ice melting in the Northern and Southern Hemispheres.

Geology

Geology is a fascinating area of research and study. If seeking or planning to seek employment in this market you will be happy to hear that the number of job openings exceed the number of students graduating with an interest in geosciences.

At a high level one might describe the study of Geology as follows:

> Geology is the study of the Earth, the materials of which it is made, the structure of those materials, and the processes acting upon them. This includes the study of organisms that have inhabited our planet. An important part of geology is to obtain the knowledge of how Earth's materials, structures, processes and organisms have changed over time.

Some of the areas of study today include soil science, volcanic activity, the evolution of the Earth's crust (especially important along the Rim of Fire in the Pacific Ocean), hydrology and some aspects of oceanography. The latter category has been focused recently on understanding the effects on ecosystems from sediment transportation over land through air and water.

The importance of studying the history of our planet is so that one can forecast events and processes that might have a major impact in the future. In addition, through historic analyses we can determine how past events and processes have affected our planet today.

Here are a few examples of responsibilities to help provide some insight into the world of a geologist:

The Study of Earth's Processes. Many activities such as landslides, earthquakes, floods and volcanic eruptions pose a threat to people. Geologists work to understand these processes well enough to avoid building important structures where they might be damaged. For example, if geologists can prepare maps of areas that have flooded in the past they can prepare maps of areas that might be flooded in the future. These maps can be used to guide the development of communities and determine where flood protection or flood insurance is needed.

The Study of Earth's Materials. We use materials from Earth every day such as oil that is produced from wells, metals which are produced from mining operations, and water that has been drawn from streams or underground sources. Geologists conduct studies that locate beds of rock that contain important metals, help plan the layout of the mine that will extract these elements and the methods used to remove the metals from the rocks. They do similar work to locate and produce oil, natural gas and groundwater. Oil deposits are often done through detonating explosives and looking at the inverse scattering (namely the acoustic wave from the explosion once it hits an underground pocket of oil will scatter back to the surface) data. Geophysicists develop analytical models to unravel this data and determine the location of the oil, size of the pocket, amount of natural gas that surrounds the oil, type of soil and rock that surround the location of interest, etc. It is really an exciting industry.

The Study of Earth's History. Today the concern is about climate change. Many geologists are studying the past climates of Earth in an attempt to understand the change that has occurred over time. This includes many sources available such as coring down into the polar ice caps and looking at the trapped carbon dioxide along with other gases in ice that was formed 700,000 years ago. By studying all the sources of information available it is hoped to understand better how the current climate is changing and how this will affect our planet and ecosystem.

Side Note: On Friday March 11, 2011 around 1446 Japan time, I was in a business meeting in a coastal town about 4-5 hours south of Sendai where the Fukushima power plant is located. We would learn later that it was a 9.0 Earthquake as measured on the Richter scale which set those now historic events into motion. At the time of the quake our smart phones lit up and we were forced to evacuate. One of my closest friends was driving home from a customer site heading North of Tokyo towards Sendai. Suddenly, as he recalled to me, the road began to "wave" violently up and down. The waves on the road were extremely violent and powerful. He told me all traffic had to come to a halt as there was no choice since no one could drive. His heart was pounding as he watched the infrastructure around him crack and crumble. The waves on the turnpike looked like those you would see if you shook a rope. His and other cars were tossed and bounced only to be tossed again. It lasted over six minutes and after it stopped, most of the traffic was completely turned around facing the wrong direction on the road. North of them in Sendai, the first wave of the Tsunami hit land and was measured over 30 feet in height. This was followed by a surge of water which swallowed up Sendai as it swelled to a height of over 60m (more than 130 feet high). I have driven thru Sendai for customer visits prior to the Tsunami and observed that it is a very flat agricultural area extending for miles. The Tsunami travelled at a speed of about 380mph as it came ashore swallowing up land, crushing houses and even a wall built to protect the town against the worst possible Tsunami the Japanese could envision. The magnitude of this event was beyond imagination. Later, from satellite geosurveys, it was determined that the Earthquake moved Japan closer to North America by almost 8 feet!

Physical Chemistry vs. Chemical Physics

These two branches of science are often confused and there are key differences that should be recognized as you look and apply for jobs. Let's get a brief look at each so the differences at least become apparent.

Physical chemistry was first coined and applied in the 1750s by Russian scientist Mikhail Lomonosov. It is the study of particulate, sub-atomic and atomic groups up to macroscopic phenomena in chemical systems. The investigations focus on the laws and concepts that govern these processes by applying classical principles and concepts of physics. This would typically include motion, characterizing various energy states, forces, time involved for a process, thermodynamics, statistical mechanics and dynamics. Physical chemistry is predominantly a macroscopic science as the majority of principles from which Lomonosov founded the science were based on the study of bulk properties rather on the molecular or atomic structures involved (this would happen later when Quantum Physics was discovered by Planck in 1900).

Chemical physics commonly probe the structure and dynamics of ions, free radicals, polymers and molecules. The subjects of interest utilize the principles of quantum theory as it applies to chemical interactions and energy flow. Experimental and theoretical chemical physics seek to understand the chemical structures and reactions at a quantum level to unmask the structure of the gas phase of ions and radicals as well as develop approximations (the calculations as you can imagine are quite complex and require a great deal of computational power) so that different phenomena can be more easily understood and studied quickly at a high level. This is important for companies as they look at different processes for new products as it helps guide their decision for investment.

3.3   Occupational Outlook

From the data published and available from the US Department of Labor and Bureau of Labor Statistics here are a few areas that are in various growth phases (Ref. US Dept. of Labor). Some of the projected occupations that are expected to grow in the coming years are as follows. The format is the field followed by the expected growth in the next 5 years.

Biomedical engineers: 67%
Event planners: 44%
Technical interpreters and translators: 40%
Audiologists: 44%
Medical scientists: 35%
Cost estimators: 35%

This is a high level look of course but indicates there are some very rapidly growing jobs that could employ technically skilled people. To be a little more specific let's look at some occupational growth rates for the Bachelor's level as available in the literature. Each occupation has an expected median salary depending on responsibilities assumed. Let's take a look at some of the data available for the BS degree.

Bachelor Degree

Jobs with 10-19% growth rate
1. Accountants/auditors $55k - $75k
2. Computer and Information system managers $75k
3. Construction manager $75k
4. Management analysts >$75k

5. Technical school teachers $40-50k

Jobs with 20-29% growth rate
   1. Civil Engineers >$75k
   2. Computer system analysts >$75k
   3. Medical Service Managers >$75k
   4. Personal financial advisors $50-75k
   5. Software developers >$75k

Jobs with 30% or more growth rate
   1. Market research analysts $75k

In addition, wireless mobile technologies is a $1Trillion dollar industry in today's market world-wide and per the Wireless Telecom Group it is expected (as of writing this book) to grow at about 80% over the next 5 years. This includes phones, hand-held computers or tablets, GPS navigation, web browser software, cameras, and so on. It is an absolutely booming industry and will continue for sometime. It is one of the reasons so much time above was spent on the semiconductor review as this drives the wireless market.

There are typically three different groups that work in the wireless industry which are:
   1. software designers
   2. hardware designers and
   3. plastics designers.
The first two are rather obvious in their functions but the third might be interesting to learn more about. The plastics designer has to come up with a CAD profile for a product that feels good to hold, looks attractive, good size, etc. The software guys need to build in the capabilities and are implemented by the hardware group. In addition to this the hardware group has to do it within a size limit determined by marketing and also the weight of the phone (who wants to walk around with a brick for a phone – although the early phones in the 1990s seem like that now – I should know I carried one internationally for years and had no

idea that someday my precious phone which I treasured would be viewed as a brick!).

The health care industry is growing very rapidly and with the retirement of the Baby Boomers happening it will grow for several more decades. An example in the ultrasonic probe market was discussed above and this product line along with wireless technology needed at medical facilities is flourishing today.

3.4   Personality Types

There is no "ideal" job that suits everyone. We each have different likes and dislikes, wants and needs, etc. But there is an ideal job for you and it takes some research in the marketplace as well as soul searching to find that "niche" where you will be happiest. Here are some guidelines that you might find helpful to consider when trying to choose the correct career path:

- Look forward to going to work
- Feel that your contribution is appreciated
- Have pride in your work as you describe it to others (ok, in a defense job you will have to be a bit clever to meet security guidelines but this can actually be fun ☺)
- Respect the people you work for and with as well as enjoy being with them
- Feel optimistic about your future with the team and the company.

When you find the right job it actually complements your personality and will enhance your life. This is a very powerful and satisfying feeling. What is interesting as you read further is to note once you begin your career, you will find more out about yourself and begin to see growth opportunities within the department, division and corporation where you work. How you combine your talents with corporate needs is certainly a challenge and is very enjoyable as it is the main combination that results in career advancement.

But finding the right job to start with is challenging. Some find that the motivation to go to work is to make money and that is their number one goal. Others may want to search for job security (this is a tough one to find in today's economy...more in "Lay Offs" below), or an emphasis on social skills, or detailed analytical responsibilities, or perhaps a very structured environment. There are three basic levels in selecting a job that meets your "needs" which are:

1. your abilities,
2. your values and
3. your interests.

These are the key ingredients that help define your "needs" in selecting that perfect career. One method that is being successfully applied to define your interests is in the realm of what is called your "personality type". For years people's interests were defined in a manner that was rather high level and not very specific although it was recognized as an important ingredient to overall job satisfaction as well as enjoying life. Today's studies involving the human brain have advanced our understanding to a new level.

The emphasis today is on classifying and understanding the different personalities which in turn allow us to better understand someone's underlying motivations, emotional states, analytical approaches, etc. It is an old science (in fact can be traced back thousands of years to Greece and India) that has been resurrected and significantly improved in recent years. Application of this science is seen to be applied by those trying to improve communication and relationships within social structures (families, e.g.), businesses (sales, manufacturing, customers, engineering, etc.), educators as well as with animals, health care workers (better assist artistic people, etc.) and the list goes on.

The brain is an interesting part of the human body to study. A questionnaire was developed in the 1920s by Katharine Cook Briggs and Isabel Briggs Myers that was an extrapolation of work published by Carl Jung (Ref. Jung) in 1921. It became known as the Myers-Briggs Type Indicator (MBTI). Jung had hypothesized there were four psychological functions basically through which we as humans experience the world. These functions are

1. sensation,
2. intuition,
3. thinking and

4. feeling.

In addition, Jung theorized that only one of these four was dominant most of the time. He also postulated that for each of these four characteristics there exist extrovert and introvert orientations for a total of eight characteristics. The extrovert and introvert concepts came from what he termed "rational" and "irrational" and these cognitive functions he termed "Judgment" and "Perception". These two pairs of functions then became associated with what people used to evaluate input data and make a decision:

1. Means of Rational (Judgment) is based on thinking and feeling from above and
2. Means of Irrational (Perception) is based on sensation and intuition.

In addition, Jung proposed that each person is more proficient at using one of the four listed functions above compared to the other three. This would then be the dominant trait of that individual.

Consider the process used to make a decision (judgment) by a person. Both the feeling and thinking functions are invoked based on the input data received from the perception function that uses the process of sensation and intuition. The person who uses thinking as a means of developing a decision prefers to view the input data from a detached standpoint allowing for what seems to be a logical, consistent process given a set of rules. Those who prefer to use feeling tend to make a decision by empathizing with the situation. In doing so these people attempt to project themselves into the situation where they feel more closely connected and can weigh the various possible solutions. The decision this person comes to is based on what is believed to be in harmony with the needs of most of the people involved. This is what they consider to be the "truth".

The MBTI model developed two acronyms given by:

1. ESTJ: extravert (E), sensing (S), thinking (T), judgment (J)
2. INFP: introvert (I), intuition (N), feeling (F), perception (P)

From this they developed 16 personality types they believe that people fit into. Today there are tests offered (some free as referenced at the end and some that charge a fee but are far more extensive in evaluation). Briefly these sixteen types are (Ref. Quenk)

1. ISTJ: The Duty Fulfiller. Serious, quiet and interested in security and peaceful living. Very thorough, responsible and dependable. Well developed powers of concentration and interested in supporting the establishment. Organized and hard working these people push steadily towards goals they identified. This allows them to accomplish a task once they focus.

2. ISTP: The Mechanic. Quiet and reserved but interested in how and why things work. Very mechanically oriented. They live for the moment and willing to take risks such as extreme sports. Uncomplicated in their needs and desires. Loyalty is high on their value list and not overly concerned with respecting rules if seen to obstruct what they want to accomplish. Analytical and excel at finding solutions to practical problems.

3. ISFJ: The Nurturer. Quiet, kind and conscientious so can be depended on to follow through. Usually puts needs of others above their own needs. Practical minded and very stable they value tradition and security. Exhibit a well-developed sense of space and functionality. Very observant about others and extremely perceptive of other's feelings.

4. ISFP: The Artist. Quiet but serious, sensitive and kind. Avoids conflict. Well developed senses and appreciates beauty. Not a leader. Flexible and open-minded. Very creative and enjoys the present moment.

5. INFJ: The Protector. Sensitive, strong, original thinker. Will stay with assignment until complete. Intuitive about people and their feelings. Strong value system which is strictly adhered to. Individualistic and not a follower.

6. INTJ: The Scientist. Independent, analytical and determined. Strong capability to turn ideas into solid plans

of action. Values knowledge and structure as well as competent people. Long range thinker with high standards. Expect high standards from others. Will follow others if others are trusted.

7. INFP: The Idealist. Idealistic, quiet and reflective. Serves humanity. Highly developed value system and lives in accordance with this system. Very loyal. Easy going unless their value system is threatened. Talented writers typically and mentally quick to see possibilities. Wants to help and understand others.

8. INTP: The Thinker. Logical, creative thinkers. Individualistic and no interest to be a leader or follower. Gets excited over theories and concepts. Very capable to turn theories into clear understandable paths.

9. ESTJ: The Guardian. Practical and organized. Likely to be athletic. No interest in abstraction unless there is a practical application. Has clear vision of how they think things should be. Likes to be in charge and very organized at running activities. Value security and peace.

10. ESTP: The Doer. Friendly and action oriented. Focus is on immediate results and lives in the present moment. Risk takers living fast paced life style. Long explanations frustrate him. Loyal to peers. Wants to get things done so does not respect rules if they get in the way.

11. ESFP: The Performer. Fun loving and enjoys making life fun for others through sharing their enjoyment. Live for the moment and love new experiences. Dislike theory but interested in serving others. Likes being center of attention in social circles. Common sense well developed.

12. ESFJ: The Caregiver. Popular, conscientious. Tend to put the needs of others over their own. Strong sense of responsibility. Values traditions and security. Enjoys helping others. Needs positive reinforcement for self-confidence building.

13. ENFP: The Inspirer. Creative and idealistic. Will accomplish anything of interest. Has great people skills. Lives life in accordance to their inner values. Excited

about new ideas but details are boring. Broad range of interest and abilities. Flexible and open minded.

14. ENFJ: The Giver. Popular with outstanding people skills. Concern for how others feel and think. Not a loner type, prefers to be with people. Very effective as managers of people and group leaders. Likes to help others.

15. ENTP: The Visionary. Creative and intellectually quick. Skills range over a broad range of areas. Enjoys debates and besting the competition. Excited about new ideas and projects but tend to ignore routine aspects of life. Can apply logic to find solutions. Enjoys people.

16. ENTJ: The Executive. Assertive and outspoken. Born to lead. Excellent ability to understand difficult organizational problems and capable to create solutions. Well informed and excel at public speaking. Values knowledge and little patience with inefficient and disorganization of others.

The idea of understanding personality type is to help one better understand themselves, what motivates them, who they think, what makes them happy, etc. so choosing a work environment that meets these criteria has the highest chance of being that "ideal" job. This information once available will help in a focused approach to job hunting and asking very detailed questions during the interview process.

3.5 Thoughts on Which Degree Level is Right for You

The question that we all wrestle with at some time is just what degree level should one pursue before going to work? Do we stop after obtaining the Bachelor's? Master's? how about going straight through to the Doctorate and then get a job? Or do we stop at the Bachelor's or Master's degree knowing that the company you will work for most likely will provide for you to go back to school and get paid as you pursue your next degree(s)? These are important questions and having some insight helps one to make the right decision. So here are some thoughts on this topic to consider.

3.5.1 Degree vs. Career Considerations

Let's take a look at what some of the career responsibilities might be for each degree received. Please keep in mind this is a high level discussion that is based on what many actually do in industry today with these degrees. It is designed to give you some idea that might help guide you in making your choice.

A Bachelor's degree gives one a set of tools that can be used to solve problems. (Actually, this is a true statement for all degrees but since the Bachelors degree does not provide the opportunity to specialize it is essential to note this here.) This person is typically trained to address problems that the company needs solved. It could be working on the bench testing products, in a manufacturing photolithographic lab, assembly area or out debugging product at a customer's site. The role is very diverse and is one value that a Bachelor grad has – she will get to see a wide range of technology and have the opportunity to move from role to role as openings are created. The employees in this group are absolutely essential to the success of the company. Without employees in this group product would not ship, costs would be too high and companies would fail.

I've been on numerous flights to Asia where I met engineers (these could be science grads or engineering grads but once in industry most everyone is referred to as an engineer of some sort) with Bachelor degrees who were tasked to go on-site at a major customer and service the equipment either for scheduled regular maintenance, training customer personnel on using new equipment or perhaps to debug, correct the problem and bring back on-line. Often they would be on-site for several weeks at a time and in some cases longer.

Master degree grads have some very exciting opportunities as they might be assigned to manage groups consisting of technical school graduates, Associate degrees and Bachelor degrees. They might be in charge of a lab like photolithography, computer software development, vacuum technology, conducting R&D testing, material strength testing, prototype development, medical systems integration and on and on. Master degree grads will find opportunities for management positions or not depending on the interest of the person (viz. your interest). They often work directly for a PhD or someone of high level in the corporate world depending on their interests and responsibilities. It is an exciting area and can offer many advancement opportunities.

The PhD is very focused in a specific area of expertise and drives new product development quite often. Depending on their interest, they can take on management roles or not based on their career objectives. Quite often they will work with the companies' marketing groups to help guide the capabilities to align with market needs (namely, don't build a square peg if the market round pegs). They often define the company's core competencies so that marketing can align this when searching for opportunities for future growth. Without this team of people, a company would be like a ship without a rudder – no technology direction to pursue. More on this in the "Chief Engineer" category below.

3.5.2 Go Straight Through for Your Degree(s) or Not?

To continue or not to continue …. that is the question! This decision is one of the important personal decisions you will find that you have to make. Should you decide on what degree to ultimately pursue (BS, MS or PhD) as you enter the university or stop at the BS or MS and later let the company you work for fund your costs for the next degree? This is really personal in choice but can also be affected by the prevailing financial climate. I've seen colleagues of mine make their decision as a result of both options and, as with everything there is a Yin and Yang that need to be considered. Here are some thoughts that might be helpful when making your choice:

- Becoming employed in industry while remaining interested in pursuing higher degree(s) allows one to better understand how a company works and what paths are available. Then tailor your degree pursuit to that path.
- Deciding to go back to school once you are working will be a strain as you will need to juggle family life, professional responsibilities and achieve a certain grade in each course in order to be reimbursed. But it does cut your costs as a student.
- Some grads worry about accumulating more debt once they graduate and opt to go into industry rather than continue in school. This choice provides a level of comfort to some as they begin to pay off loans and make some money which improves their lifestyle.
- On the other hand, graduate schools often extend teaching assistantships or (usually later on when the advisor feels the student is ready) a research assistantship. These are not much in terms of income but do offset expenses so that you do not accumulate additional debt if managed carefully.
- If you stay in school to continue pursuing advanced degrees you may even find that industry has invested in a particular technology within your department that funds student involvement. This not only gets you introduced to the sponsoring corporation (and vice versa) but gives you a chance to get acclimated to industrial work. Not every university has this and not every department offers this so do your homework if you want to seek this as an option.
- Some have stayed in school as the financial climate at the time of graduation was not secure and available jobs were unattractive.
- Popular combinations to combine with technical degrees while working are MS/MBA, PhD/MBA or if you already have a Bachelor's degree, you could go back to school to obtain the MBA. This course work would be approved in the company if you express an interest to move into the business side of the company and it fit within the company needs.

4.   HOW DO YOU START?

Once people learned of my theoretical physics background and that I was a math junkie, I was often asked if I find math hard. Math seems to scare many people. But when I'm asked if I find it hard, I usually tell them "No, math can be quite simple" and give them an example like in Figure 6.

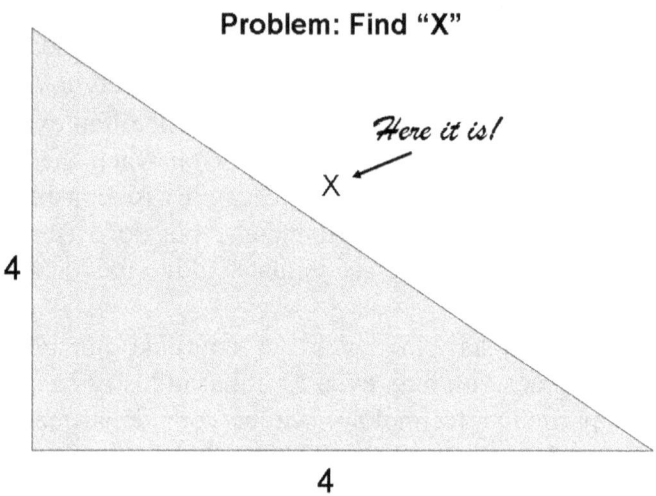

Figure 6. Find X

So you can see, it is all perspective and "hard" is relative. If you think finding a job is hard…I say, nawww….keep it fun as hard is all in how one perceives things.

It is important to keep in mind that when you start you will find your work will be challenging. What you will also experience is that when you go to staff meetings you will begin to see there is so much more going on within the company and most likely you won't understand what is being said initially. This is normal so don't get bothered. You can not enter a university by going direct to your senior year – you know that now. You have to build up to it and it is the same in the company you will work for. But you may feel that your contribution seems to be a small role in the company and the program you are assigned to support. That would be a mistake. Suppose you worked in the automotive industry and found a way to build a part quicker than the procedure that you were taught. So, knowing time is money in a company, you implement and begin cranking out the parts. The danger is that without qualifying the new procedure you chose to use the integrity of the part is unknown. This could jeopardize safety. Instead, it might be better to point out your idea and solicit approval from management before implementing. Or perhaps you go to work for a firm that services DoD programs. And assume that a technician was assigned to solder special lumped elements onto a circuit board. These positions are usually filled by graduates of a 1-2 year tech school who have certification to a certain J-standard soldering level. Assume the tech is soldering a timing sensor, for example, that goes onboard a missile. Once the missile is fired, the timing sensor wakes up the circuit which allows the missile to receive course corrections from the launching platform so it lands with more precision. But if that solder joint is 'cold' so that there is a high impedance or maybe cracks that occur due to the high-g launch the timing mechanism doesn't perform its function. As a result, the missile can not be corrected and may miss the target – or worse hit a friendly target. The war fighter's life, which we support through DoD programs, depends on everyone doing their best job no matter what part of the product they support. So please keep in mind…maybe you think your role is small but the war fighters do not and they thank you dearly as you help save them from harm.

As you look at large companies from outside, it can be intimidating. You'll find that you'll ask yourself questions like, "How do I fit in?" "What is the corporate structure"? Without understanding something about corporate structure, it makes it difficult to comprehend where and in what capacity you might be hired. So to help provide some sense of understanding, let's take a look at a typical organization chart as shown in Figure 7 below.

There are many types of company organization schemes but this is a typical one that you will often see. Let's take a look at this and work our way through it so you get an idea how and where you might fit in.

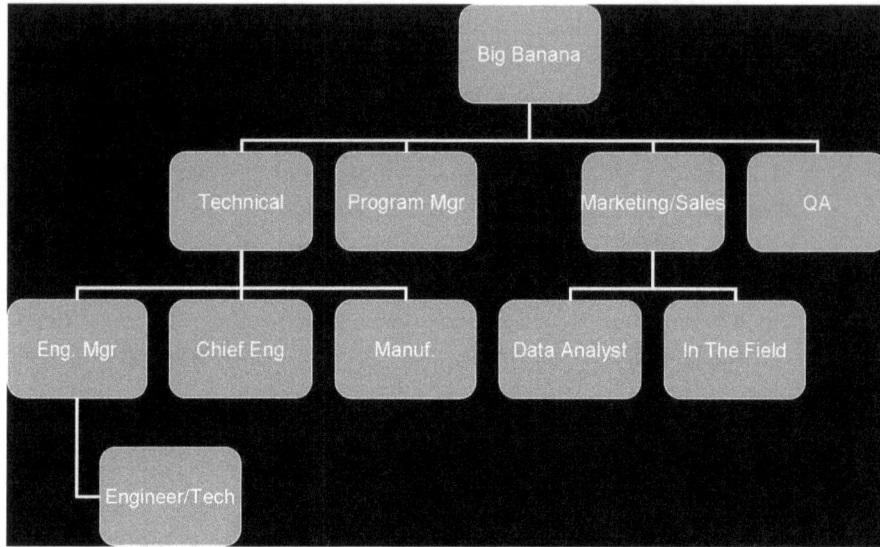

Figure 7. Org chart

1. The Big Banana. This is the top person who has overall P&L (Profit and Loss) responsibility for his division or department. He could be the company president if in a small/mid-sized company or in larger companies it would be a VP (vice president) who reports to an Executive VP, who then reports to the President and then to the CEO. In fact in the large OEMs, there could be three

Big Bananas, one each for the Technical Division, Program Office and Sales/Marketing department. These three then would report to an Executive VP who reports to the President and up to the CEO. But whatever the scheme, the Big Banana here is the boss of all that goes on under his/her group. This is typically divided up into four categories. So let's look at each separately and not in any order of importance per se.

2. Quality Assurance (QA). This is an independent group within the company and is tasked with ensuring that all work done in manufacturing and engineering comply with the required work standards, such as those listed in the MIL-STD handbook (Military Standards – a set of tests and assembly protocols required for military products) or other industrial work practice standard called out by the customer. They roam the floors, do random testing, support costs for Qualification tests at the beginning of a new program and several other functions.

3. Marketing and Sales. Usually, marketing is the group that makes initial contact with the customer and identifies the work that is needed. It may involve an initial proof-of-concept (POC) design to see if it is feasible or not and if so gets an initial order and turns it over to the program manager (PM). The PM would then work with Engineering and the customer to complete the POC. If this goes well the next step would be that Sales generates an order for what might be called a qualification phase. Again, this would be turned over to the PM to go through the development phase and qualify the design to the customer's needs. If a space application is the end product the design must meet certain MIL-STD (military standards) so when up in space there is a low risk of failure. It is not easy to repair once in space so the Qual-set that engineering builds (a small quantity that is to be qualified to survive and function by subjecting it to many extreme conditions) must represent all future products so when samples are tested randomly during production, there is confidence that they will meet the same Qual-standards.

After the Qual-run as it is called, a pre-production order is generally awarded and this is followed by production runs, which could go on for years. The production runs are where companies make their money, so again you can see the importance of marketing. Marketers need to estimate up front the quantities for production so companies can estimate the return on investment (ROI). If the ROI is too small, there needs to be a decision whether to support and enter the market or not. There are a number of factors that go into this decision, but this is another important and exciting aspect of doing the statistical analyses on market estimations that we will not address here.

Within the Sales and Marketing group are those out "in the field" and the data analysts. Those out in the field are the ones that are very closely connected to the customer. They either travel frequently to stay close to the customer (personal relations build trust which is important in business) or live nearby and visit often. The data analysis group is focused on analyzing data presented to them by marketing people. An entry level here would be either in the field or in data analysis.

4. Program Management. The PM works with everyone involved in filling the purchase order requirements from the customer. This can be working with engineering to layout a milestone and schedule with deadlines to be met (ensures on-time delivery and staying within costs), interacting with the customer to keep them informed and also with Sales. This can be a very detailed job and is certainly fast paced. They are also the ones who provide any bad news (and good news as infrequent as that is) to the customer. So working closely with the customer, getting to know your counterpart at the customer site and staying close to engineering and manufacturing within your company is very important. If this is where you want to eventually go...my suggestion is to stay on top of the job as a PM and do not let there be any surprises to you and especially to the customer. Alerting the customer to problems can sometimes be a robust discussion internally within your company and can be politically driven. But it is a dynamic, exciting position. There maybe some assistants to

support the PM at larger companies, otherwise they work alone. This is not an entry level position but one that you can work towards if you find it interesting.

5. Technical. There are typically three groups that fit in this category as illustrated in Figure 7, which are R&D, engineering design/development and manufacturing. The Chief engineer usually focuses on R&D which is driven by the road map from the Technical Marketing group. In a smaller company, this person may also be the key designer on a difficult design and hence gets integrated into meetings with customers. In larger companies the Chief engineer is a career path that is just technical. This person is solely focused on technology that is relevant to the companies' growth and can be very lucrative. The person filling this position is usually deemed as an expert in the field and is not an entry level position you would initially apply for in job seeking.

The Engineering Manager (EM) wears many hats. One important aspect is that this position requires estimating hours expected to complete new anticipated work. Two other important responsibilities include (1) managing the engineers to make sure their projects are on-time and within the budget developed during the quoting phase and (2) laying out R&D study phases based on the marketing roadmaps. So, if Sales received a RFQ (Request for Quotation) from their customer, the EM would estimate the number of hours necessary to complete the work required. It is an important line-item element in the company's quotation to the customer since if a problem occurs and more time is needed, the engineers/scientists are among the highest paid groups within the company so profits can be eaten up very quickly by these cost overruns. As you can guess, in that case the Big Banana would not be very happy! Also working with manufacturing to ensure that all goes smoothly is a key aspect since the company can lose considerable money very quickly if there are problems when going to production.

Manufacturing is another area that is in the Technical section and typically separate from Engineering groups. The responsibilities here include hours estimated for production runs for sales to obtain prices for quotes, sometimes purchasing is in this group or would be a separate group in larger companies. Purchasing needs to place orders for raw materials in a timely manner and leverage quantities of orders to obtain the best price. (If McDonalds didn't order enough hamburgers for the holiday rush and ran out....there goes profit!) Manufacturing also schedules work for production so product is shipped in a timely manner to the customer. If shipments are missed and the customer has a production line shut down because they do not have your product, their Big Banana calls your Big Banana and guess what? So you see how this is working? The drive is profit!

So this is it...everyone has an important role to assume and all contribute to the bottom line of the company's P&L. The higher up you go in the Org Chart, the more responsibility you have and the more impact on the bottom line you will have. This increased responsibility is met with some very lucrative perks and a high salary with a nice bonus plan. But remember the earlier discussion chasing the dollar may not be what is best for you. An honest review of your interests will dictate if you should pursue this career of corporate growth in the long term or not.

Again, this org chart is very general and each company is specific unto its own structure. But the general functions will be as illustrated and the basic functions of each block will be as indicated above.

Now in terms of job selection if you are that lucky to receive numerous attractive offers, here are some thoughts you may want to consider.

1. **First, Prioritize.** Some of the parameters (ahh, now that is the scientist coming out in me to choose that word) which should be considered in this analysis, but not limited to, are:
   a. Job description,

b. Location,
c. Salary,
d. Benefits
e. Do you have a significant other? Will this person also work and is the location right?
f. If you plan on children, can you locate within the community where you want your children to go to school?

2. **Don't be too picky.** There is no perfect job. Find happiness in your responsibilities, lifestyle and where you would like to go in your corporate career.

3. **Research the company/division/program**
   a. How long has the company been in business?
   b. How many employees work at this location?
      i. If a large company, how many are in the program you will be assigned to support?
   c. Is the program big/small in terms of engineers? Has it been financed for long? What is the probability that funding will continue?
   d. How does the program fit into the market (e.g. commercial or DOD end use)?
   e. Are you happy with your future co-workers? Your future boss?

4.1   Job Hiring Breakdown

The data that is presented in this section is the courtesy of AIP Statistical Research Center (www.aip.org/statistics) and is noted in the References Section. Now that we have discussed a typical organization chart for a corporation and possible career paths that might be interesting to consider as you gain experience, let's look at the different fields where new hires are typically employed and the respective salaries of each. We'll begin with Figure 8 that illustrates the breakdown on the fields of employment for Bachelor's Degrees for Physics (Engineering data is shown in Figure 9 and can be seen to be comparable to that of Physics -- more detail is listed on the IEEE.org website). What is interesting to note (and will be seen as a common trend in Physics in the following figures) is that half of the Physicists hired go into the private sector that consists of using engineering related and computer based skills. These engineering related skills can vary greatly but would typically consist of bench-related work like programming automatic test stations, studying the design-assembly process, generating masks for photolithography, IC circuit layout, correlating test data, etc.

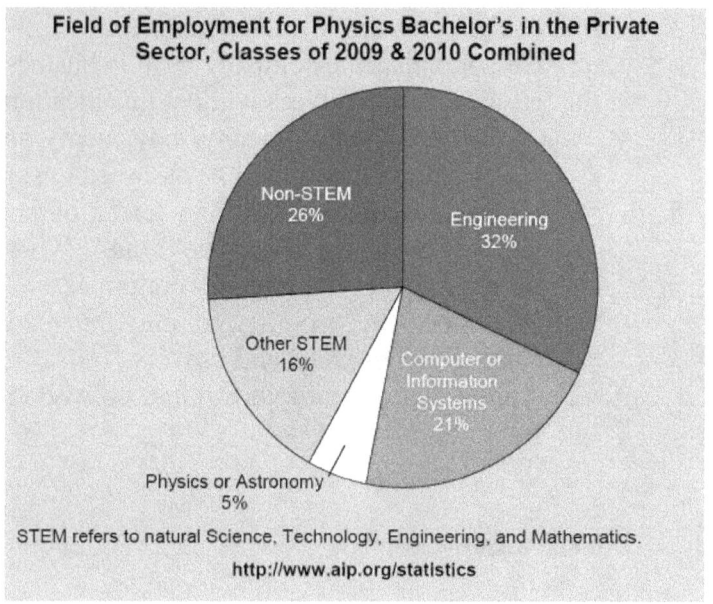

Figure 8. BS Physics Employment

The salaries associated with the BS in Physics and other disciplines are illustrated in Figure 9 for 2009 data (the latest available as of this writing).

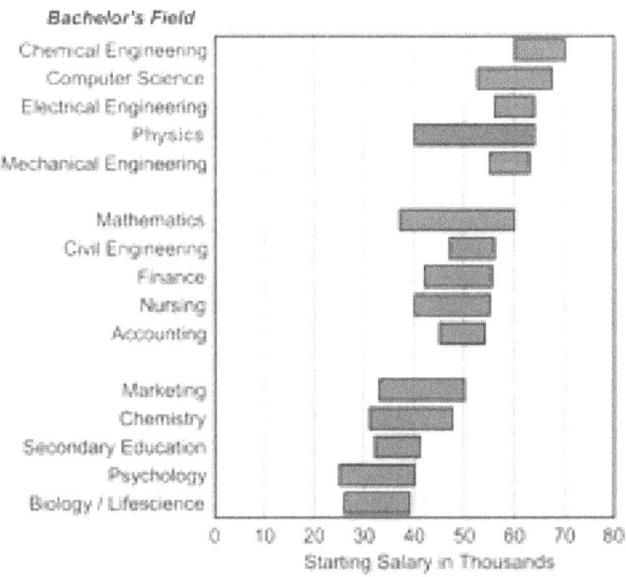

Figure 9. Starting Salaries for BS Degrees
(Courtesy of www.aip.org/statistics)

Note the wide range in salaries for BS in Physics compared to most other disciplines displayed. The reason is that Physics related class work prepares the student for applying the learned skills across a very wide range of employment opportunities as seen in Figure 8 above. Each employment sector has different salary requirements it can support, hence the wide range in salaries.

Side Note. In order to obtain an estimate on current starting salaries using the data in Figure 9, assume an average for inflation of 3% / year. For example, suppose you wanted to know the starting high salary for Chemical Engineering in 2012 to see how it compares to the value listed of $70k in 2009 from Figure 9. Take the difference of 2012 − 2010 = 2. Why 2010, because inflation is already included in the 2009 data for that year. Then multiply (1.03)*(1.03)*$70k to get the inflation adjusted income for 2012. This is a very good means for estimating current salaries and is used often in industry to even adjust product unit prices.

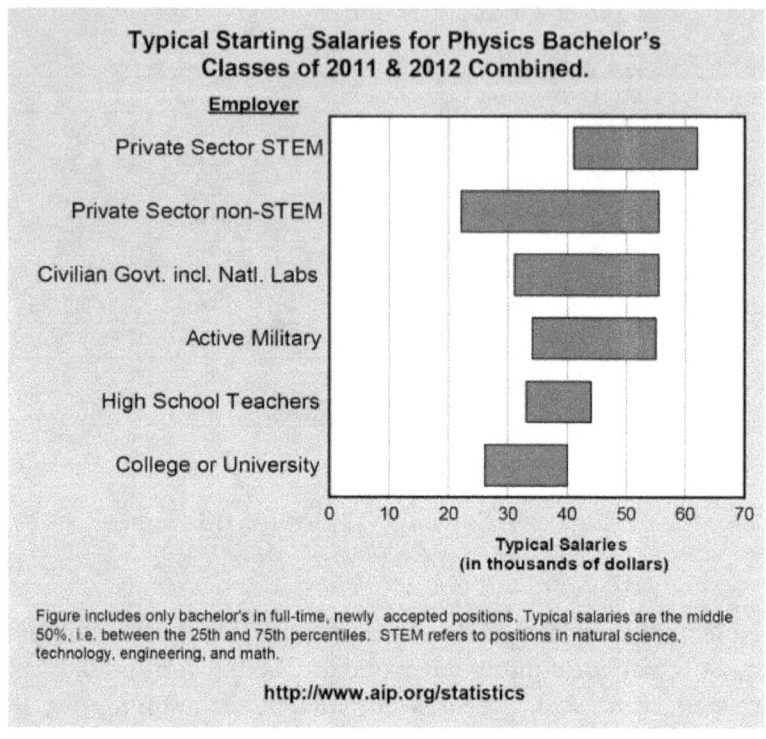

Figure 10. Bachelor's Starting Salaries in Physics

From Figure 10, Bachelor's Starting Salaries in Physics and comparing to the range shown in Figure 9, Starting Salaries for BS Degrees, one can see there has not been a significant change. As I write this book I polled my friends in the private sector who are in a hiring capacity about this trend. The feedback received was that this trend basically continues today but there has been a slight up-tick in salaries during 2014's fiscal year.

For Master of Science graduates, let's look at the details in the employment breakdown as seen across several years. Figure 11 was compiled using data from 2006, 2007 and 2008 whereas Figure 12 used the latest data available by averaging the results for years 2012, 2013 and 2014. Comparing Figures 11 and 12 one can see that employment is up in the private sector, civilian government jobs and military service. Employment is down for employment at universities and high schools from previous years. One important commonality that should be noted is that about 50% of the hiring occurs in the Private Sector and this continues to grow in today's economy.

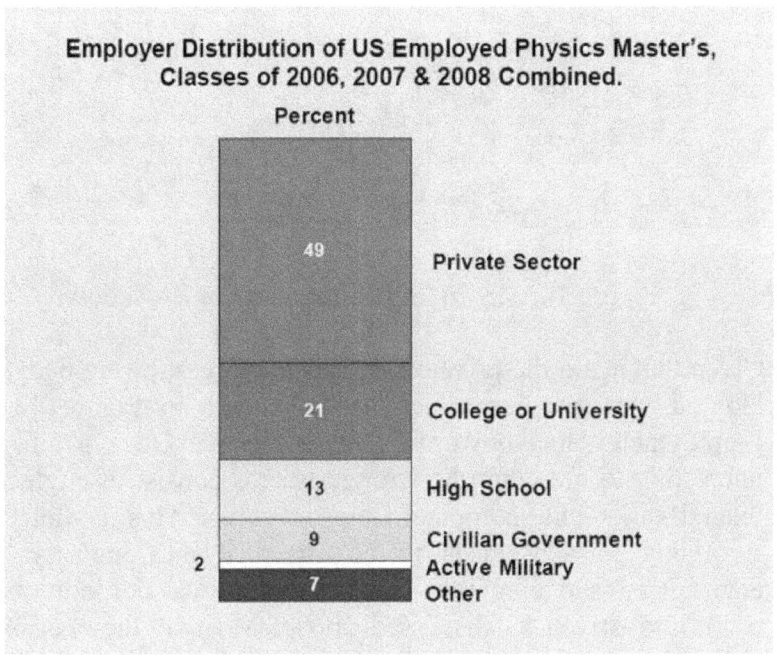

Figure 11. MS 2006-08 Employment Breakdown
(Courtesy of www.aip.org/statistics)

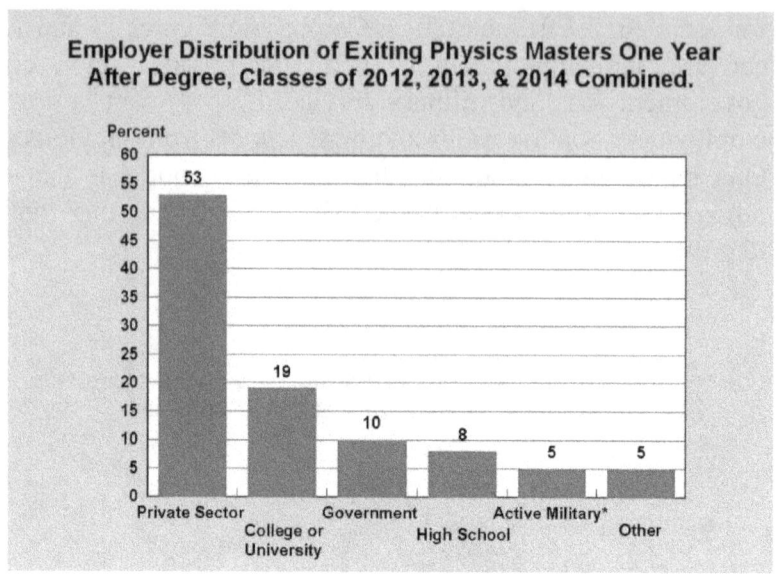

Figure 12. MS 2012-14 Employment Breakdown

Given that again the private sector is the largest hiring body, let's look at the specifics which are displayed in Figure 13, MS Employment Breakdown in Private Sector. Once again, it is interesting to note that the strongest areas consist of engineering related tasks and computer science skills. This is not really surprising for Physicists to be absorbed into industry using engineering and computer related skills since corporations are profit/loss driven as discussed above. Most of these efforts in industry are not searching for the next quantum theory unless it relates to a product that can be built and sold. If that were the case then there would be opportunities for more pure research. Instead, industries support the development at universities for R&D in the pure science but basically use theories that are already accepted within the technical community which can lead to product development. Overwhelmingly it can be noticed that the category of "Engineering" dominates the field. This does not mean it is filled with engineering students. Everyone who joins the private sector will get a title of "Engineer" since the ultimate goal is to design, build and ship. So physicists, biologists, mathematicians, etc. will all become "engineers" in the corporate sense and is true no matter if you have a BS, MS or PhD.

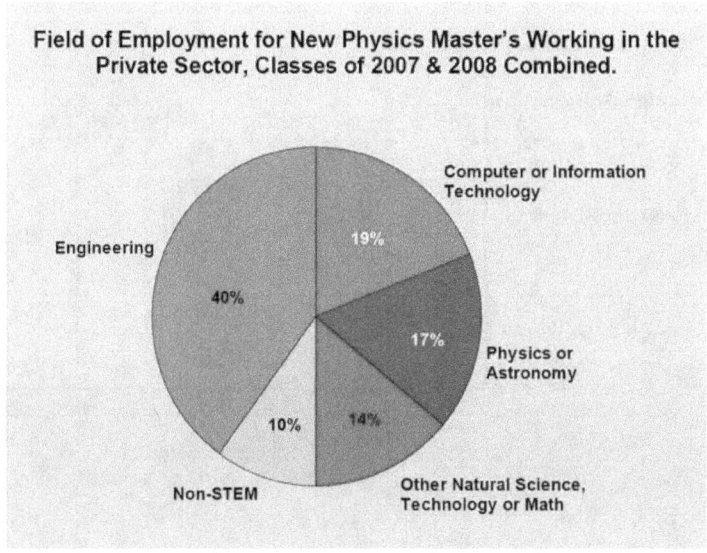

Figure 13. MS Employment Breakdown Private Sector
(Courtesy of www.aip.org/statistics)

Salary wise, MS degrees in Physics break down as illustrated in Figure 14 using data from 2006, 2007 and 2008. This represents the latest starting salary information available at the time of this writing.

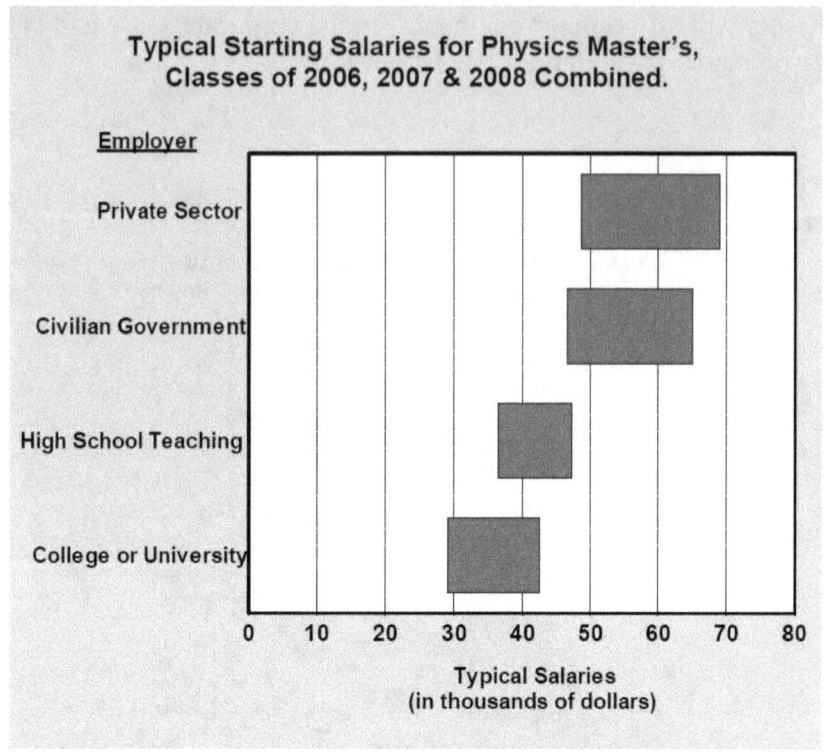

Figure 14. MS Salaries per Employment Sector
(Courtesy of www.aip.org/statistics)

Another interesting trend that comes from examining the above salary graphs is that since the private sector makes a great deal of money, it can afford to pay their employees better than other employers. It is important, to reiterate a point made earlier, that one define her job priority and it is not recommended that one only chase the high salary. If you chose a job that you really don't like just to make more money, in a few months the thrill of the check wears off and you are left to contend with the negative emotion of dragging yourself to work everyday. That is a very bad feeling! So define your job and life priorities when searching and making a job/location selection.

In Figure 15 one can see the salary breakdown per employer for PhDs in Physics (and is again comparable to that of engineers).

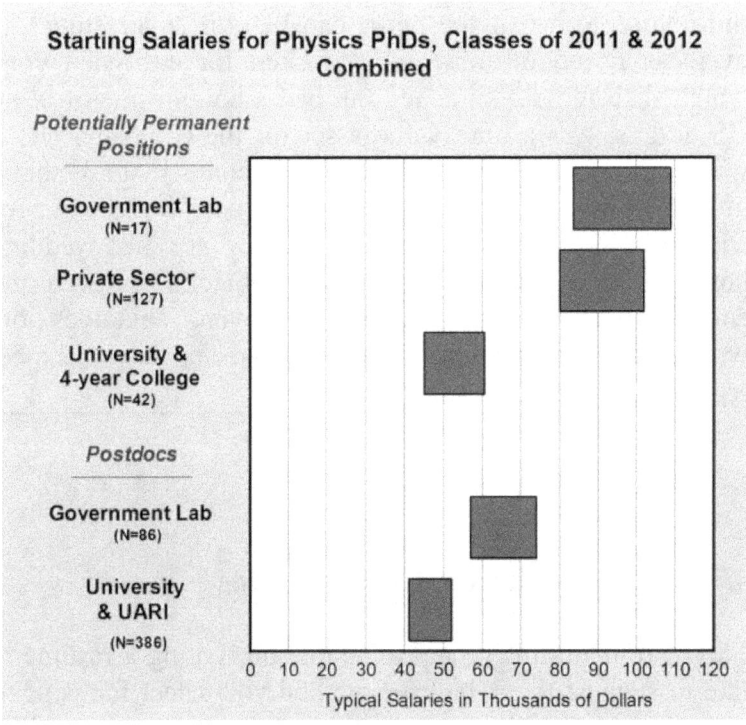

Figure 15. PhD Physics Starting Salaries
(Courtesy of www.aip.org/statistics)

What is important to note is that these are starting salaries. Why is that important? As the private sector is focused on P&L and the PhD will have a significant contribution to this for a company, the reward in terms of salary increases rapidly. As the employee develops a reputation within the company and his contribution to the "bottom line" of the P&L sheet is seen, bonuses and salary and other "perks" increase substantially. A company does not want to lose a substantial contributor to their profits as everyone makes more money and reaps these benefits. Today government labs aggressively compete with the private sector for PhDs (and other degrees as well) but in the long run, those that enter the private sector will benefit more financially.

Side Note. Some of the perks can be quite interesting. A close overseas friend of mine who worked for an Asian company, which shall remain unnamed but had a bottom line of billions of US dollars, was a high value asset for the company. His design work was critical for his division of the company to remain on top in the wireless product market. When he announced his engagement the company paid entirely for his wedding and honeymoon to a destination of their choice. They even provided consultants to help organize the entire event. This does not come without a price tag on my friend of course as he was expected to remain very dedicated to the company.

4.2   A Suggestion on Resume Writing

There are many references available on writing a resume but I'd like to suggest that whatever method you select for your resume you consider adding a short paragraph at the top as discussed next. Keep the body of your resume the same which will include education, work experience, etc. At the top of the resume under your name write a short paragraph about the kind of job you seek. Tailor this to the job description you are applying for and use what you believe to be the "key" words from the company's job description in this 3-5 line paragraph. Tailor this paragraph for each application you submit. In some cases at the large corporations, there may be a scanner looking for those "key" words in your resume. If it finds these "key" words in your paragraph it passes your resume onto a human to review. In order to make your resume stand out when the human reads it, tie these "key" words to your job experience or class work. Don't expect them to spend time trying to do that. They have too many resumes to review. So help them out. If they see that the key words they are looking for can be tied to your experience, your resume will get circulated to the hiring group(s). Then you stand a very good chance of being asked in for an interview.

Remember, they are the ones doing the hiring, not you. So the short paragraph topping your resume should be focused on their needs and not your skills. There is a subtle difference. Your first "job" is to connect the two of them, viz. connect your skills to their needs as best you can albeit they only provided a short job description when posting it. This 3-5 line paragraph has two effects besides getting you past the robotic interpreter:

1. The hiring agency will see that you are conscientious in that you made the effort to try to connect your experience to their needs.

2. It will get you to think and prepare in a concise manner how to respond if you get an invite for an interview (see the discussion below on "Elevator Speech" in the Interview Section). You will most likely be asked how your background could help meet their needs. Human Resources will often ask "What type of job are you looking for?" or "Tell me a little about yourself" and now you have that answer already prepared!

Suggestion. If you are asked "Tell me a little about yourself" be cautious on this. It is sometimes used to solicit information that companies can not ask in order to find out if you are a single Mom (hence may want more time off from work), how old you are, etc. Stay focused on remaining professional and what you need to do to get the job.

****

<u>**Example**</u>:

YOUR NAME
ADDRESS HERE
CONTACT INFO

Structure a 3 line summary of the job you are searching for. Pick "key" words from the job description posted by the hiring agency. Modify this for each job you apply.

EXPERIENCE

<u>**DATE**</u>   <u>**DESCRIPTION**</u>

Keep all the info here in the body of your resume the same for each application. But make sure it is complete.

****

5. DAVID VS. GOLIATH (SMALL VS. LARGE CORPORATIONS)

As with anything, there are pros and cons. So you probably are not surprised to find that this also applies when choosing between a small vs. a large corporation. To help guide you in making this choice, one needs to get an idea of some of the key elements in each category for each type of company. This is the objective of this section. So let's begin with

5.1   The Small Corporation.

Let's start by discussing some of the Pros in the small company that you may want to consider when making your career choice. Small companies significantly dominate employment here in the USA, which is a very interesting statistic and available from the US Census Bureau.

PROs

1. Wear more hats. This means that you will have more responsibilities in a small corp. which can be quite exciting. As a result, you learn more about a wider variety of responsibilities. For example, you may be working on a design for one customer in your office when you get a call from the testing lab that the measurements on a product you designed for another customer did not meet spec and they need your help. So you can see that you will experience a great deal of variety in a small company.

2. Highly dynamic. This can occur in many varieties. For example, the customer may request early delivery of a product in order to satisfy his customer. So you will need to become very creative to accomplish this. Or perhaps your raw materials are received late and it is a challenge to meet your deadlines. But when you meet the deadline and

receive a nice fat bonus check, you will sit back and think life is so good!

3. Multi-tasking. Typically you will be assigned a number of responsibilities for products that are ordered from a number of customers. So, each product most likely will have a different end use. As you meet and brief the customer on her job status you learn more about the end uses which is quite exciting.

4. P&L resp. The Profit and Loss will eventually fall on your shoulders as you gain more experience. The Engineering Manager will solicit your input when quoting a new job and will ask your inputs on such items as

   a. how long do you think a new design received from sales would take based on past experience, or

   b. do you believe there would be a need for any special, expensive materials that need to be identified up front to bill the customer when quoting.

   This extra responsibility of course comes with a higher salary and bonus. This responsibility is assumed after you gain experience.

5. Independence. In a small company there are fewer people and fewer yet experts in different fields to solicit input from when facing problems. In addition, there are fewer people (that is why it is a small company) to access for support. So basically, since there are less people, this results in providing more diversity in your job role in order to complete the design for the customer. It requires a willingness to accept different responsibilities as needed to meet customer deadlines and stay within cost. This feeds back into #1, #2 and #3 above.

6. High visibility. As you gain more responsibility, you gain more visibility by upper managers. Reputation begins to develop. Keep this in mind as you begin working so you develop a positive image. Face problems with confidence, work well with others under stress, do your best to meet delivery, etc. All these are important in developing a solid reputation.

CONs

1. Lower Tech expertise. Since you spend a great deal of time wearing many different hats in a small company (see #1 in Pros above) you do not spend as much time researching a specific area of interest. So when you face a problem, you look for a solution quickly to stay on schedule and within the budgeted cost of the project. You will learn a great deal about many different areas which can be very exciting, but you will not, most likely, become an expert in any one area.
2. Smaller programs. As a small company, they will receive orders from bigger companies (your customers) which will entail less work to complete. As you do complete one project you will move quickly into another program and will not learn as much about how your product will be used.
3. Volatile budgets. Large companies can cut future demand very quickly and in many cases without warning. Since small companies do not have deep pockets this can be challenging for upper managers to keep everyone employed.
4. DOD – black programs. Working on these programs will require a US security clearance. And it will require that you do not discuss your work outside the company with your friends. As did I, you will get used to this lifestyle as you accept the significance of what you are doing. And one other note, given the level of clearance, you most likely will not have any idea how the product is used. Several years ago, I consulted with the US Navy on one threat of great concern to the US. I had to visit a facility to brief the customer and Navy on my results. I arrived and went through several armed security check points. I was finally escorted by an armed guard and took the elevator several floors below city streets. I had been in this facility

many times but never knew there were offices below ground level. There was a labyrinth of halls that I was led through, all painted the same color as were the doors. There were no distinguishing marks. Finally I arrived at a door and after announcing my arrival, my armed escort handed my over to another escort who brought me to the briefing room. The area was outfitted with some very sophisticated electronics at the time and was quite an adventure to experience.

5. Growth limited. In small companies, there are only a few manager level positions and this can impede growth up the corporate chain. If you like to acquire titles this could be an issue. If you are an important contributor to the company, it will be recognized and in some cases special positions are provided to help keep the employee happy and give that person a chance to grow. But this is rare.

6. High visibility. This is a double edged sword as there are some pros as you saw above. However, if you miss deliveries or overrun your budgeted costs it will become recognized by upper managers.

5.2   The Large Corporation.

These companies are the movers and shakers in the global economy. They can appear clumsy and bureaucratic from the outside when compared to the high dynamics of a small company, but make no mistake, they are bottom line driven and so all the managers are tasked with reducing costs and increasing profits. It is demanded by the share holders and this flows down through all levels of the corporation. Without these Goliaths there would be no Navy ships, no satellites, no wireless infrastructure, no medical advancements, etc. It is phenomenal to witness first hand what can be accomplished from within these large corporations. So let's discuss some of the Pros and Cons of joining a large firm.

PROs.

1. Very senior engineer/scientist position. The large firms recognize technical excellence by offering positions where the employee just concentrates on a specific area of technical expertise. They could become very specialized and an expert on a particular product. They may just focus on a specific technology and research this area. (Note: In some of the overseas OEMs and here in the US government defense labs, this person could have little customer interaction and often would not be assigned any task with design requirements including deadlines. This does not happen in today's economy as much as it used to – but there are still a few around in the US.) So how does a position like this get justified? The expectation is that through their expertise the company might develop a technical edge to provide a better solution when bidding on a government RFQ. If successful, the win is in the hundreds of millions of dollars and up. So the stakes are high as winning a large contract will keep tens of thousands of people employed for many years.

2. Large programs. As noted above, winning a large contract is very lucrative. But funding is never guaranteed in today's global economy. In some cases in the defense sector if the program being funded is deemed to be absolutely essential to the security of the United States and its NATO allies, it may remain more secure as overseas governments would participate in the finances as well as the US. But whether it is for defense or not, a large program means there will be many people involved. In general there is less opportunity for an employee to wear different hats as in the small companies. (There is an exception to this but depends on the individual – please see the Side Note below). It is true that deadlines for completing certain milestones on large programs are often longer than in small companies, but there is also a lot more work in between milestones. The pressure can seem to be less if observed from the outside but this would be

false as there are many more tasks to be accomplished on a large program and milestones always seem to come up quickly. And because these programs are so large, program/product development can span different federal government administrations, so cuts can happen quickly and are often deep. This will affect not only the OEM through massive lay offs but it will flow down to the smaller sub-contracting companies as well.

3. Future growth. Although there are additional higher level positions available in large companies vs. small corporations, competition is higher. But if you are successful at entering into the management role, and are successful as a manager, upward mobility is encouraged.

4. Location transfers. As there are many divisions within a large company and each division tasked with different responsibilities, the opportunity is available for the employees to request assignments to other groups. This can be very stimulating as these large companies have divisions throughout the US and abroad. So getting a 2-3 year assignment in the UK or some country of choice could be very exciting if you like the idea of experiencing other cultures. On the other hand, if you are interested in moving to a different technology that is the focus of a different division, that has merit as you do not lose any seniority through the move and keep your position/title as well as salary.

5. Large budgets. This provides a more stable work environment in many cases as funding is quite often secured for a number of years. In addition, profits are higher and as you grow professionally many perks become available. Having worked for a large firm I always flew First Class and often had drivers meet me at airports and hotels in foreign countries. It was a tremendous experience – however, I generated a great deal of highly profitable sales for my manager and division so this is a positive example of what high visibility can do with perks extended.

> **Side Note**. Often in times of crunch (to get a product out the door to meet deadlines and avoid fines or perhaps to finish a major proposal before the RFQ expires) companies will develop what are called Tiger Teams. If an individual has the initiative to get involved one then gets a great education on numerous items within the company. Special treatment is provided to these team members (equipment, services, resources, fast response on "hot" items, etc.) as well as recognition if all goes well. It is fast paced and exciting not to mention you get to meet some of the best talent within the company when getting involved.

CONs

1. One of many engineers. Initially, you will be one of many in a large company. There are those who have considerable experience in a wide range of technologies that you can solicit information from, but to gain recognition and grow professionally is more challenging as the "playing field" is larger.
2. Growth can be slow. Although you can draw on a large range of experience from #1 above, this does make growing professionally more challenging in other areas. There are more people competing for new positions that may open within the company that you may want to apply for. This will certainly make change and growth more challenging. Also, this is where bureaucracy can be challenging, there are many rules that help foster growth but can seem to get in the way of one's interest.
3. Less independence. As a member of a large team, your responsibilities will be highly integrated within the various team members. Working closely with others can

be demanding on occasions; on other occasions it can be very rewarding. The "kudos" for a good job done that management bestows will be to the team and not an individual.

4. Only 1 task. As many people are involved with the program, your responsibilities will be focused on a specific task. There will be very little opportunity to cross-over and learn other disciplines without applying for a transfer.

5. Low dynamics. Given the long milestones and deliverables, the work environment is typically not as fast paced as in small companies. In small companies you will be assigned a number of projects for different customers and will feel that you are being drawn and quartered at times. So in the large company, the work environment is challenging but with fewer dynamics.

Personally, I have enjoyed working in both small and large corporations where revenues were from $15M/year up to $20B/year. So the initial choice of where to work should be one that fits your personality, career needs (as you see them now because if you look back at my career highlights, you will note that objectives can change significantly) and what your heart tells you.

5.3   Collaboration between Universities and Industries

There are many values for University departments to be tied to industry. Two very important benefits are the ability to place students upon graduation as well as soliciting funding for future growth within the department. But let's take a closer look at how this can be a win-win situation for both industry and the University.

1. Introduce students to the work environment. By obtaining funding from a company, students can be employed within their own university department. This gives them the opportunity to work on a task that is of interest to industry and for them to get introduced to the industrial environment as there will be milestones that will need to be met – albeit these milestones are "soft" as education comes first. But they do exist and helps to train one to think in this manner. It is also consistent with the training one gets to meet deadlines to turn in work assignments for school classes so the work environment mirrors and supports the educational process.
2. Companies' benefits include
   a. Being seen as supporting education, which is a prime concern for the US to grow.
   b. Putting profits back into society is always seen as positive as well as a tax incentive.
   c. Obtaining results from a university using their resources to develop a necessary tool or to investigate the feasibility of an idea reduces corporate costs. So investigating specific issues this way is quite inexpensive compared to committing resources from within the company to perform the same tasks.
3. Program sponsor can evaluate. As the company funds a project within a university's department, there will be reviews of the work as well as the student(s) performing the work. This can be a tremendous asset for several reasons:
   a. The student may be asked to participate on a summer program at the company. Here the student gets further introduced into the industrial work environment and will meet a number of engineers who already work there.
   b. The company can evaluate the student to see if there is a fit within the company upon graduation.

c. Training at a lower cost level within the university occurs so if the company does hire the student less time needs to be spent developing the technical skills needed to perform a job.

d. The student gets an opportunity to see how class work is applied to solve problems in real world product development.

4. Higher funding brings more good students. As the university receives more funding, additional resources can be purchased (test sets, computers, etc.) which contributes to developing a high tech facility. This will beget growth and growth brings in very good students and funding.

5. Additional placement service opportunities

a. Jobs can be "created" based on performance at the university. By this I mean that a company can create a job for the student worker if there is good synergy.

b. Presentations at corporate conferences provide visibility for both the university and student. It is also a very good learning experience for the student to be involved and one that the student can find very rewarding. Today in order to sell yourself (including the interview process which we'll discuss below) and your work you need to be able to present well.

So from the above, one can see that Physics, Engineering and other sciences should reach out or continue to reach out to industry for funding some level of work in which students can participate. Small companies will typically draw on local talent as it becomes available from local colleges and universities. The small companies typically will not hire as often as a large firm (hence they are small) and not having the working capital of larger companies will not seek candidates across the country due to the expense of interviewing them as well as relocation. The upshot of this is that small companies will tend to hire locally.

Large companies however, will look globally for solutions to their hiring needs. As a result they have an interest in seeing good talent develop wherever available. Many divisions even have further extensive training programs to finer tune the capabilities of their new hires. For example, many students today are coming out of school with good backgrounds in digital technology. However, in military electronics, there is still a large need for RF (Radio Frequency) design – albeit beginning to wane. In fact, I know of one large OEM that has a training program on RF engineering for those who will be designing and laying out printed circuit boards. This capability is an art as well as a science. But it is necessary until it is completely replaced by digital circuitry.

So there is a complimentary goal among universities and industries that needs to continue developing. Universities want to produce talent that is useful to society. And industry wants to hire and use this talent. It therefore makes sense to define some common ground between both so each supports the other. This commonality could be enhanced from the funding by industry for programs/projects within a university that students can be a part of. The programs/projects would be those of interest to the funding agency, here a specific industry. The universities have well established and proven core programs and through these funded efforts could help transition students to industry. It would provide a means for students (and faculty) to consider various methods to apply the theory taught in their curriculums. Applications lead to product which leads to sales and this is a motivation for industry.

From this effort industry gets to fund the training of students at a much reduced rate compared to what it could do internally. In addition the sponsor would have the opportunity to evaluate students and determine if any or all would be a good fit to the corporation.

The students get indoctrinated into an industrial setting through working on this sponsored work. In addition the university would have an opportunity to place more students better qualified for the work force through this training process and in terms leads to more funding efforts as well as an increase in student enrollment. This is a win-win situation.

5.4   Thoughts on Career Advancement

As you start your career most likely you will be overwhelmed with everything going on around you and all the information you accumulate as you learn your job. It will feel like your brain is on overload. Don't worry it is normal. If you did not feel like this you probably would not be employed long. It is not meant to be some corporate hazing ritual to new employees but in fact there is just a lot of stuff going on in companies. Don't lose sight of your responsibility which is to learn your job – this is your priority as it is what you were hired to do. So don't deviate, stay focused. As you get comfortable at work you will find that you begin to understand more about what is going on inside. But you only see it from your limited perspective as higher up in the corporate structure there is a whole different world. Many of the very upper level management teams are not expected to know details about the technology their company is known for – but certainly know it at a high level and where it is used. You most likely will eventually know the details of the product far better than these upper level managers as they were hired for an entirely different set of skills.

While you begin to get comfortable with your job there will be career paths within the company which will become available. It will serve to further your understanding of possibilities and help motivate you as you plan your career. Or perhaps you will be happy doing what you are doing for a long time as you will grow in seniority and responsibilities. But any advancements from this career choice will eventually end and you will need to choose either to continue pursuing what you are currently doing (which will basically be "flat") or make a move and change rolls. There are many who do not pursue advancing their career which requires they accept new roles within the company. This can be for many reasons but most of the time the choice is made due to comfort level. The person is content with his job, income, family life and outside interests and his current role within the company allows for a very nice balance. So he is happy. This is perfectly fine – what can be better than being happy and making a comfortable living surrounded by friends and family?

But there are others who are interested in shaking up their status quo and going for different career paths. The big steps in salary, bonuses and perks along with corporate titles are exciting to pursue. Some people will seem to be on a fast track as they get promoted quickly. These people eventually hit a limit as their knowledge base was never developed. It takes time and a great deal of hard work to develop knowledge and capability. Think about it, have you ever seen anyone graduate from high school and enter a university only to get a PhD in one semester? You know it is impossible. But you will meet or observe people who go up the corporate ladder quickly since they learned how to "play the game" as it is called. There are many ways to fast track a career but it is not my preferred method nor is it really respected among that person's "peers" and certainly not from the working levels below this person's position. For this reason, we will not discuss any insight into some of the ways it has been accomplished in the past, which I've witnessed. Instead, let's discuss some of the ways you can enhance your career as a young engineer which will gain respect among your colleagues as well as generate self-respect and self-confidence.

1. Understand the tools needed for your job. This means if you work with equipment, learn it inside and out. This applies whether it is a network analyzer or CAD tools. Knowing your tool kit that is necessary to do your job also

means you will find the limitations. This is invaluable to your manager when capital equipment or software is ordered as you will be in a position to help suggest what should be included that will help make your (and others) response quicker to complete work assignments. Remember, time = money!

2. Learn your profession. This does not refer to just your job (although again, learning your job is where you need to start) but learning more about what happens in the product flow before and after you in the company. If you don't you will basically be a robot; but as you learn more about your profession it makes you more valuable to the company as you will be cross trained on other integrated disciplines. This cross training will allow you to more easily move from one area of the company to another using the same skill set – this is very valuable to you. If at some point funding in one area is reduced, you can move to another more profitable section. As you become more knowledgeable more people will seek out your advice, which will open doors for you with the opportunity for advancement.

3. Management. If you enter this area you will have a great influence on the people working for you. Get to know each of their skills. What do they like to do? Encourage them. Delegate – this shows you have respect for their abilities. Do they have any problems you can help with? Respect them and always stand up for them when times get rough. If mistakes get made dissect the problems with them but emphasize the positive so they continue working for you and not against you. Encourage them to find the solution. Maintain the protocol – they are basically the horses pulling the cart and you are the driver. Don't try to be a horse with them.

4. You will have success. Remember Hans Solo's comment to Luke Skywalker in the first Star Wars movie, "Don't get cocky kid"? Be careful you don't fall into this pit when you have some success. Enjoy your moment when it comes as it was the result of hard work that will garner

you admiration among the team. A very close Japanese friend of mine told me of a popular saying in Japan, "An empty can makes the most noise" ... I like that saying.

5. You will make mistakes. My Grandpa once told me when I was growing up "You learn from your mistakes". He thought for a moment and then added with a laugh, "So I (Grandpa) must be a genius". ☺ You will make mistakes but don't dwell on them as they could bring you down. Same philosophy when doing something well – dwelling on that can bring you down. Acknowledge it (never make excuses), admit it if necessary (you don't need to send out an email announcing it!) and dissect it. What led to the mistake? Figure this out and replay the scene in your mind visualizing your solution. End this mental imagery as you visualize yourself successfully solving the problem. (Some of the very high profile sports figures use this exact method after losing a game – it works!)

5.5   Lay Offs

The focus of this book *Road to Success* is not only to help identify some of the expanding opportunities in industry along with how to proceed in applying for jobs but also to introduce the reader to corporate life and what can be expected. Lay offs are unfortunately one of those unpleasant elements that exist in pursuing a professional industrial career and everyone must recognize this. Of course no one wants to experience this but it can happen and quite likely you will experience this once or twice in your career. Don't fear being laid off but prepare for it so you can minimize the impact. To understand the process better let's look at how it can occur and then we'll discuss some thoughts based on personal experience (yes, I've been laid off in my career) to help prepare in case you do experience a lay off.

It is quite natural that once you are laid off you feel as thought it was a personal attack on you by the company you worked so hard to support. Well it is not personal but is in fact a sign of a struggling and weak portion of the company. For whatever reason, the program or division that you work within begins to struggle financially. This weakens the company and upper level managers see this as a cancer growing (in debt that is) which needs to be cut out in order for the remainder of the company not to be significantly affected. Remember, companies are all about profits. Hence, lay offs. In some dramatic cases, buildings are closed and the real estate is sold. In other cases a "skeleton" crew is maintained in anticipation there could be a slow recovery back to health. If you are laid off, recognize the fact that the company has some very large financial problems and this kills any opportunities for professional growth for its employees. Consider it a blessing in fact that you were laid off so you can re-group, redirect your focus and get another job where you can continue to grow and learn. Your professional career and family life will be much healthier. There is always clearing after a storm.

The actual process of being laid off is of course not pleasant. In some cases I've seen companies post a list in the cafeteria and if your name is on it you then report to HR (Human Resources) for exit papers, then to your desk to clean out your personal effects, and finally escorted out the door. Sometimes the company will provide a severance package or perhaps set up a room with computers for job searching or offer job re-training – but this is not very often the case any longer. It does happen but not often. In other situations your manager will give the bad news personally to you, then HR escorts you to your desk, clean out your personal items and then out the door. There is no time for good-byes. The process is very sterile.

To reiterate, a lay off is not personal even though you will most likely feel this way at first. Your thoughts will wander to what will happen to you now; the morning after feels strange not having a job; what are you supposed to do as this was never a course offered at the university. Many thoughts will go through your mind. So here are some suggestions that you might want to consider in preparation of the possibility that you are laid off.

1. It is estimated that it could take 8 – 12 months of job searching, as this book is being written, to find another job. So while you are working it might be a good idea to make sure you have liquid assets for at least 8 months to live on based on your current monthly spend rate. You will find this cash cushion will last longer than 8 months as you will tend to cut back on your daily spending. No more going out for lunch, you will make your own morning coffee, etc. But think of this…you lived on a strict budget in college and you found ways to be happy. So you know you can go back to these roots which were established in school if necessary.

2. Work might have pulled you away from your family more often than you wanted so take the time to reconnect with them as you also search for a job. Find ways of having simple family fun. You will all grow stronger as you pull together and developing a tough mind set that you will come out better from this experience will go a long way to successfully going forward. Work at maintaining a positive attitude. Do not let negative thinking linger in your mind – it will drag you down.

3. Collect unemployment. Some feel strange doing this but it is your money. That's right, on each pay stub there is a line item where the state's unemployment was collected from you. So collecting unemployment as you job search is helpful to pay some of the bills and again, it is your money. It is not a hand out as some feel, it is basically a savings account that the state collected from your weekly pay and is available to you when laid off. HR will give you an official letter of termination and the reason for termination – lay off. Use this when you sign up to collect.

4. Be persistent. Set aside a certain number of hours each day and work intensely during this time to find opportunities and apply for them. Also, don't forget to follow-up so the company will note your interest level. I can assure you, if you do not give up, remain flexible in

job and location, you will come out stronger than ever. Your commitment, resolve, determination and family ties will all strengthen. So hang in there....when you are seemingly at the bottom you can only go up but you have to be the one to climb.

6.   OUR NATIONAL SECURITY

As noted earlier, this book not only intends to help one find and secure a professional technical job but also what life in industry entails. Security is one element that transcends all boundaries – private sector, government and universities are all affected. Unfortunately, in today's world, cyber intelligence and espionage is becoming an important part of our lives. To counter these threats, we need to become educated on the latest means used to unwittingly gain information that is used by hostile governments and people. In a word, we need to remain vigilant. Besides the much publicized hacking of credit cards and personal accounts at Big Box stores, overseas governments target:

1. the Defense Sector:
    a. government funded military programs
    b. Space programs related to defense
2. the Commercial Sector:
    a. Industrial developments
    b. Commercial Space programs and
    c. University R&D.

Let's look at each of these areas as security has become an important part of life in today's world. We will begin with the concerns regarding the US Department of Defense programs.

Department of Defense. This section includes defense related work in space programs as well as those programs at various industries and universities. Most of my career was spent working on defense related programs. As I gained experience, I began to travel internationally to NATO friendly countries to understand more about the threats these governments faced and present technical methods available to address these concerns. So it was easy to find information about me professionally on the internet. Since I had been integrally involved with several very high profile military programs, not only did I get more requests overseas to give talks but also became highly visible to groups subversive to the US government. One example came when I received an email that immediately sent up red flags when I read it. First, here's some background that you'll need to understand on this particular case.

When a company builds a product that is classified, it is given a classified part number and serial number which is labeled on the product housing or package. The product is then shipped via classified carrier to the customer who integrates it into their system and this then is sent out to the field to be used by our War Fighters. It could be a missile, radar system or any one of a variety of products. This particular product under discussion here where I was involved was not visible to the public as it was highly classified and no one had any access to the product description except only a few select people within the military and our customer. It was not to be spoken of outside the company I worked for except within the confines of a secure facility with the select group of people who had clearance to know about it. The email I received had this part number, serial number and some of the functionality listed about it. That was impossible for anyone to know outside our group unless the enemy had actually obtained it and physically took it apart to perform some rough measurements. The email went on to ask if I would please provide more information about the product as they would like to place a large order. I immediately called the FBI who is tasked with managing the counter terrorist threat at home and told them what happened. An agent was instantly dispatched to read the email and my computer, not being classified, had to be 'scrubbed' of any evidence of this email, especially to make sure it was no longer available as I travelled globally. However, working closely with the FBI I acted as an interface with the sender of the email. From this we learned where the email originated (country and location actually) and gained insight into the group responsible. The overseas group eventually decided there were too many questions and "disappeared". Namely, they left no trace of the initial "company" that generated the email to me but from this correspondence they were indentified here at home so they could be monitored more carefully. And it was noted they remained very active! If they actually had learned what they wanted about our product, they could have developed a counter measure to render our military system using this product totally ineffective. This would have been a loss of billions of US dollars as well as putting our war fighter's lives at risk as it is part of a very expensive product for the US Military. If you do go to work for a defense related company, I highly recommend you take security very seriously. It may seem silly in some cases and from the position you hold, but ask a war fighter who is out in the field dodging enemy fire how she feels about security if you want an honest answer.

And if you would like some interesting reading on a very clever, overseas scheme of electronic espionage, there is a declassified report now available called the Mandiant Report (Ref. Mandiant).

Commercial Areas of Concern. This includes commercial satellites, industrial product development and university research that would have a commercial application. Many programs at universities are at the fore-front of leading edge technology as is often the case for large corporations. Usually corporations have security measures in place to minimize cyber attacks as they know if they lose their edge on competition by losing their next generation product from an overseas spying operation, their bottom line will suffer and may in the extreme case, cause them to go out of business. Universities, on the other hand, are still quite relaxed when it comes to guarding against cyber intelligence. Since universities have always been spared from various attacks over the years, the feeling has developed that academic studies and interests are beyond the interest of spy networks. And with the pressure to publish or perish, the attitude persists as time is so limited. But it should be recognized by all that there are people overseas who look at everything. They want to either steal information or gather ideas that could be used to develop a product which could be a threat to our homeland security or to the War Fighters' lives. As James Comey, the Director of the FBI, so aptly put it, "There are two types of groups here in the US. Those that know they have been compromised by overseas spying eyes and those who do not know".

There are some obvious areas that are being watched carefully here today in the US and a short list follows. It is not meant to imply that if your work does not seem to fall into the short list noted below that it is not of interest to overseas governments but these are some areas of great concern. The following is a short list of some technologies that need to be carefully monitored for cyber espionage:

1. Power Grids. If the enemy were to capture our power facilities' control system and shut down power at will, this could disrupt industry, defense, financial markets, transportation, government, health centers, etc. It could also pose a nuclear meltdown threat if they were to gain control of the reactor core and we all know what happens as evidenced from the terrible Fukishima accident.
2. Water contamination. Urban areas rely on water which is vital for life. Agriculture uses tons of water for irrigation and requires that the water be pure so crops and livestock are not contaminated.

3. Federal Aviation Administration. Most flights today are due to business travelers. Flight Control Towers at airports present an attractive target for would be terrorists. Imagine what might happen if these control tower computer systems were taken over by a threat -- this would be devastating to our economy.
4. Home Land Security. Threats could cross borders and develop cells within the US for covert operations. If compromised, Home Land Security issues could be a major transportation concern as we all saw on 9-11. This could have a tremendous impact not only on our nation's security but on commercial economy as well. Without our strong economy it would be very difficult for any of us to prosper and live productive and happy lifestyles.
5. Leading Edge Commercial Capabilities. Stealing technological secrets where the US might have spent hundreds of millions of dollars to develop for commerce use could result in tremendous loss to the US economy. As industry is profit driven, this clearly would have an impact on a company's survivability if their products were compromised. Universities are often involved with high level R&D and this means there are people watching who intend to do harm. If a university's research is compromised, this would impact future funding, outreach programs for placing students in respectable positions and all this would have an impact on the future growth of the university. The reputation and credibility of such an institute would be severely compromised.

Side Note. Help on security issues is available from the Federal Government. Today, the FBI's top priority is counter terrorism. They have local offices that are conveniently located and given my personal experiences, I recommend that industry and universities connect with their local FBI office. In turn, they can help keep everyone informed and will be of assistance to investigate any suspicious contact that maybe received. The State Department does not have local offices placed throughout the country and are understaffed. The background check on individuals, for instance, is very superficial when compared to the depth provided by the FBI. The impact on this is if one relies on the State Department providing information to help screen overseas applicants for employment or to invite to join academic programs, they do not have the man-power and resources to cover this search in sufficient detail. The FBI I believe is the best suited government agency that can most effectively support local government, industrial and university needs at the depth necessary to provide the highest level of security. Establishing contact early by a university or corporation before there is a need would be prudent.

7.   THE INTERVIEW

So you have done your research now and found a few areas of interest along with the companies that offer job opportunities in the areas you would like to pursue. That is not an easy task to accomplish but at least you are on your way to securing employment. The next step is to get a resume out along with the necessary paperwork required by the company and wait for their response. And after a short eternity, you get some positive feedback after who knows how many rejections.

Let me comment about the rejections you will receive as we all experience this. Interviewing is like a game....some you will win and some you will lose. Accept it and note that it is not personal. Like any sports team, if you do lose review in your mind why you think you lost (sometimes if you call personnel at the company they will share info like this with you), make the necessary adjustments and hopefully this will help make you even stronger the next time.

But remaining positive with a strong outlook and continuing to send out resumes, you will eventually receive an invitation for an interview. And these letters are exciting to get. So you then go about filling out paperwork for the company and make your travel arrangements. If flying, you may find that someone will meet you at the airport or a rental car will be made available. You might also find that once at the hotel you will be invited to dinner (usually this happens for some MS and most PhDs) or maybe they will just expect to see you at their lobby at a specified time.

Allow yourself plenty of time – the last thing you want is to arrive late as some of the people you plan to visit may not be available since everything during the interview is scheduled way in advance of your arrival. These people have tight schedules to maintain. Keep in mind the overall drive is to achieve the highest P&L for each quarter, so time is money inside a company. And in addition, arriving early will help eliminate being upset from traffic or whatever by arriving late. So give yourself some time. If you would like, even make a "dry" run on the evening before your interview so you know where you have to go....but in any case, you finally arrive at the company's entrance at a good time. (Note: if you are not authorized the use of a rental car for your interview, you most likely will be met in the hotel lobby by someone or directed by the company to take a taxi. Either way, plan to be early.) If it is a secured facility, there will be guards (armed most likely so don't be alarmed) that will check a list to see if you are authorized to be on the company's campus or not. This list at the guard station is received from the Human Resource contact that you corresponded with during the time you set up your interview. But if there is a problem with paperwork at security, have your contact information readily available so security can call and verify why your name is not on their daily list of visitors. It is rare that this happens so don't let it bother you if it does happen – and by the way, this is the company's mistake not yours. They'll apologize when you meet them.

After leaving security you are directed to the parking lot of the building where you are to report. Also, in some defense related programs and companies, there could be several armed check points you will pass through on your way to the lobby. Once you arrive and park in a visitor's spot or where you were directed to park, you grab your brief case and walk to the lobby.

Figure 16. A Corporate Building

Some places will require that you press a button or call on a phone at the front door to request permission from security within to open the door. Be aware, as you drive onto the premises cameras will be watching you. Once inside you announce yourself at the front desk and sign in. They will check your briefcase and ask if you have any recording devices. Once this step is complete, take a seat as they call your contact and announce your arrival. Or if they have a small "museum" within the lobby, it is nice to look at their display while you wait as it gives you some insight into the company.

Suggestion: make sure your phone is off so it won't ring. Also while you wait do not busy yourself with your phone. Sit quietly – it shows patience and demonstrates that you are relaxed.

Suggestion: if you are in a secured facility, they will not allow any recording devices. If you forget and bring your phone (typically it has a camera) with you and they request that you check it with lobby security, leave it with your rental car keys at security. Then you will not drive off and forget the phone when you leave.

HR (Human Resources) will generally greet you at the lobby and escort you to their office. Usually this promenade includes some small talk like the weather, your trip, hotel accommodations, etc. which you could endure or you could use the time while walking to get more information and show interest in the company:

1. solicit some information about the campus you drove through to get to the lobby,
2. if the building is new inquire about how long it has been up, what work is done in this location,
3. discuss the display in the front lobby,
4. if it is a large OEM how many work at this address,
5. how long has your escort been working here (who doesn't like opining about themselves?) or
6. basically get some high level information about the company before getting into the details that will come up during your interview.

Suggestion: make good eye contact with everyone you meet, shake hands firmly (not bone crushing) and sit up straight. This shows good energy, confidence and interest.

Side Note: The Elevator Speech. Whether you are a new grad searching for a job, an experienced professional out in search of a new career or in sales the elevator speech is a necessary tool. It is called elevator speech as it is meant to be about 30sec long but captures the essence of what you want to convey. Think of it this way, if you got on the elevator with someone who could make a difference in your life and you knew when the doors opened she would be gone, what do you say to hold her attention and perhaps even be invited for further discussions that could lead to a big opportunity? Hence, the term elevator speech was invented.

If you are job hunting this should contain (1) who you are, (2) what you've done (experience or education is appropriate here) and (3) why are you the perfect candidate or why is your product the right solution (if in sales).

I typically suggest to students who are new at job hunting they use 15sec to pitch who they are and the remaining 15sec to respond on how they are the right fit. Use benefit focused terminology to help drive your point stronger. For example, "I'm an EE graduate with a strong track record of identifying and solving problems that will improve performance and cut costs" vs. "I'm an EE graduate with an interest in design work that will help support the program". See the difference? Companies want to save money, get an edge over competition with performance, quicker delivery and lower cost so the first speech is more appealing. The first example keys off the knowledge of what companies are all about ....P&L. So use benefit focused phrases.

Once you arrive at HR and get seated (they may offer coffee, tea, trip to the bathroom, etc. that you can either accept or not) they will review your itinerary for the day with you. It will have names and times on the list they will provide to you. Avoid discussing salary at this point as this is usually done following the interview. Often a company overview is provided and some of the specifics of the programs that you will interview for during the day will be reviewed. Ask any questions you would like about the day's schedule but the details you should save for the face-to-face meetings that are scheduled. After this brief meeting with HR (usually 20-30 min.) someone will escort you to your first contact. This will begin getting into the details of the potential job you seek to fill.

As you begin the interview process, here are a few questions that might help you get some additional information as well as generating a good impression on the people you meet. So a few items you may want to ask:

1. **Life 9-5. Any flexibility?** Determine what a typical day will entail so you can visualize what work will be like. Do the people all start at a specific time? Or is there flexibility in work hours? Many people you will work with may start between 0800 - 0900 so this maybe is what you will hear back from your interviewer.
2. **Specifics of job.** What exactly will you start on? What training (if applicable) will they offer? Again, get into the details so you can understand and evaluate if it is right for you.
3. **Program big picture.** What is the end use of the product? Learn a little about why this product/program you will work on is needed.
4. **Program lifecycle.** Is the job you are interviewing for an upgrade to an existing product and this product maybe quite important to national security so it goes through many upgrades? This can also be true of commercial products. Also, you would want to know if it is coming to the end of its funding cycle and what happens once it is completed.
5. **Well funded?** This is a good one to give you a sense of longevity on this program and within the company's division.
6. **Professional growth.** As you begin your career, what is a typical growth pattern you might expect in 3 years? In 5 years? (Ask this of the managers, the engineers you meet will not know nor be in a position to speculate.)
7. **Engineer/technician headcount.** How many engineers are on the program you would join and how many technicians work on it as well. This will give you an idea of how big the program is within the company. HR will most likely give you a head count of the number of people who work for the company worldwide (if applicable of course) and how many are employed at the facility you are at.
8. **Meet "your" boss.** This is important…you need to like this person as your reviews and raises are conducted by your boss. So it is good to know her and like this person. The boss (your specific Big Banana) will be important to your professional growth.

When you are invited for an interview at a company, remember that there is a real need for the position to be filled. This means they have funding in place for a new hire and also the need is almost always immediate. As programs have deadlines to meet, they need people to work on specific elements of the program in order to comply with their schedule. Often deadlines are accompanied by pay periods to your company from the customer, so it is important to the company to meet these deadlines. If your company misses deadlines there could be penalties invoked which would cost the company money. (Some of the OEMs to Wireless Providers pay several thousand US dollars a day if some of their key deadlines are not met.) So either way, deadlines are important. Those who interview you will be looking to fill a position immediately so they can get back to work on meeting schedules and get your help as well. So here are a few things to keep in mind that the company interviewers are seeking:

1. **Fill the position.** It is all about deadlines and meeting them.
2. **What are your strengths?** (know them) This is a very common question that you can expect.
3. **Will look for your weaknesses** (know them). Another very common question that will most likely be asked of you.
4. **May test you under pressure** (stay cool). Deadlines can cause stress, especially if the group you are in is behind their milestone chart to complete their deadline. So you might be tested with technical questions to see how you react. They know you feel pressure just because of the interviewing process. But they will push you to (1) really get answers they need in order to evaluate you for their team and (2) see how you react. Stay cool and do your best to answer. If they seek your opinion on how one might solve a specific problem, after you give your response ask them how they might have considered solving it. This helps develop rapport with them and they get some respect for you as a worthy team member

and colleague. In addition, you get some very good information and insight into how people think.

5. **Will you fit into their team professionally/personally?** As you most likely will work closely with others especially during your training period, it is important that they evaluate you on your ability to learn, be open to learning (namely, no ego in doing it your way or excuses, etc.) and just overall comfort level with you as a person.

6. **How can you help solve their problem?** This is a key element from their perspective. Here the interviewer is really trying to assess your capabilities for a specific problem that needs to be addressed and compare you to others.

7. **Level of enthusiasm?** This is very important. Without it you most likely will not be selected and offered a job. Too much excitement and this also is a turn-off. Asking them questions is a great way to show enthusiasm. Work related dialogue is a good thing. Also as you leave to go to another interview with another person, make sure to shake hands, thank them and make good eye contact. It is another way to show enthusiasm and respect.

You might want to consider these following ideas and how to use them in your interview. These (and your responses) can be used several times throughout the day with different people. Here are some suggestions that you might find useful to include:

1. If you were in an organization at school you might want to note this involvement demonstrated your ability to work within a team.

2. If you had a high course semester, note the level of multi-tasking this required. As you grow professionally, this will be important, not so much when you just start working.

3. The fact that you had many favorite courses demonstrates that you have interests in several areas. Be careful when and where you use this (HR is a good one where you can use this in a large company) as the team interviewing you want a specific problem solved and may not be interested in your many interests unless it can help them meet deadlines.

Namely, maybe you like writing software and also enjoy setting up test equipment for specific needs that would help meet deadlines.

4. It may be of interest if you are interviewing for a position that is not clearly tied to your background that you note how your education has helped develop your problem solving skills that can be applied to many areas.

5. Emphasize that you liked all you heard during the interview, so identify some key points to show you weren't sleeping. This is best when back at HR to wrap-up the interview.

Side Bar. I want to emphasize the importance of team effort in corporations. I had a conversation with a very high level manager of a department who told me his hiring practice is to surround himself with only exceptionally smart people – 3-sigma out on the bell IQ curve. I countered and said I would want one or perhaps a few of these exceptional scientists or engineers on my team depending on the size of the department I managed. But we need to note there are very few "Einsteins" in the world. The point of a corporation is to develop, build and ship product to make money. Most of the people I have hired have demonstrated they have good academic credentials as well as the ability to work with others. For example, suppose you choose to hire all strong people in order to accomplish a task like move a car out of the mud. So that would be the corporate objective – move the car out of the mud. However, all the independent strong men decided on different directions in order to accomplish the task and as they are independent they all pushed and pulled in opposite directions. The chances of freeing the car are slim to none. But a team would decide on a direction, join hands and together they move the car out of the mud. Now it is very nice to have a strong man or two in this group pulling with the team, but I am not a fan of all independent strong men in my department. Having discussed this with others, it appears the team concept is the preferred approach with the majority of company executives.

Once you are escorted back to HR for a wrap-up after the interview, you might want to consider exploring a few items with them.

1. **Ask for any feedback.** After you leave each person, she will call HR as well as her manager to provide feedback about you. This is done immediately while details and impressions are strongest. Your HR point of contact will begin a database on you that will be used to judge you against other candidates. HR may or may not answer you directly if you ask. But by asking it shows interest – be careful how you do this so it doesn't show desperation.

2. **Show enthusiasm for the position.** Don't rank the different projects you interviewed for as it may exclude you from the others. For example, suppose you interviewed with five different people and noted that #2 was the most interesting. However, #2 offers the job to another candidate but #3 had interest to offer a position to you until he hears that he was not your top interest. Don't rank interviews ...show enthusiasm for every position you interviewed for. Remember from our earlier discussion above, nothing is going to be perfect so look for items that interest you in all possibilities.

3. **Know a start-date and their need.** This is a good question to ask so you have a timeframe to consider in case you want to take more interviews, or maybe a vacation after graduation or whatever. But be willing to negotiate this start date keeping in mind they need to meet deadlines and that is why the position is opened. But have an idea of when you would really like to start if there were no restrictions and be ready to negotiate from there.

4. **Compensation.** This package consists of several elements and you might want to make a matrix (there I go again being a theoretical physicist) or similar method to use as a comparison of the different companies and opportunities you will need to consider when making a decision. Here are the important essentials that are grouped within compensation package.

a. Salary. Before you interview, research the salaries in your specialty by looking at the data from your professional organization, viz. AIP, IEEE, APS, etc. Also, there are websites that allow you to estimate the cost of living (such as the one listed in the Reference section below as "COL") increase as some areas of the country are more expensive to live in than others, so these cost increases can be justified. HR may even suggest that you could never receive that type of salary that you are asking for – it is their way to save money for the company and is just part of the game. But even if they don't do this, you can ask how much the job is paying. In fact you could ask that before they ask you to provide your expected salary. The trick is that if you ask too high the company may feel they can not satisfy your needs and any offer could be jeopardized. On the other hand, they may really want you and reluctantly accept your offer (I've had this happen and it is a nice feeling! But I didn't take the job as it didn't meet my priorities.) If you ask too low then it could be seen as low self-esteem, or you haven't done your homework on salary or maybe they'll be happy with that as they were prepared to offer you much more. So the key is to do your homework as best you can. This negotiation is always fun but in the end, the group that wants to hire you will have some say in salary. They most likely do not have the final say as there are company guidelines to be respected so everyone is treated equally, which are considered at the time an offer is extended.

b. Moving expenses. This should include transportation (for you and your spouse) for house hunting, temporary housing while you apartment/house search, moving and storage expenses and sometimes some help with meals.

c. Vacation, holidays, personal leave. 2 weeks is typical to start with for vacation, usually 10-12 corporate holidays are provided and there are various guidelines for personal leave.

d. 401k with matching. Find out the company policy on this and compare to other offers. You may not find this important immediately but if you plan to stay at this company for many years, it will be a significant element to consider.

e. Continuing education. If you want to pursue additional course work, ask what the policy is as most companies encourage this.

f. Benefits (medical, Long/short term disability). Here, there should be Dental, major medical, long and short term hospital stays, pregnancy (if you are the husband can you get time to help at home as your wife recovers?), baby sitting, medical deductibles (doctor visits and specialists), etc. Make a matrix and compare to other companies as this is part of your total compensation package.

Finally, there will be interviews most likely where you know within just a few minutes that the job is not for you. So what do you do? First, it is better to say "No" to their job offer than to have them say "No we don't want you". No one likes to see a rejection letter. So try to convince them it is a job for you despite your inner feelings. However, if you do get a rejection letter that surprised you when you thought everything went well, immediately call your contact and find out as many details as possible so if you did/said something wrong you can make the correction. One student in Physics recently noted to me after one of my university talks that at the wrap-up of an interview he was told he would do better if he applied to a larger company that did more sophisticated work. It surprised him as he really wanted the job he just interviewed for. What was the problem? He oversold his capabilities. Interviewing is an art and using the bio-feedback which you sense during your interviews is paramount to adjusting your responses. Both capability and need have to align for the best match.

Here's a second reason to go for the job despite knowing it is not what you want. In the job you actually do finally accept and begin working at, there will be times when you have to do things you do not want or like. That is just life. Use the time during an interview for a job that you know you will not want in order to strengthen your resolve to win and do your best despite the fact that it is not of interest. (At least this will make the interview process more interesting!) Practicing this for all occasions where you have to do something that you personally prefer not to do will develop a strong mind set to get things done despite the perceived inconvenience. This will get you back to your happy mode quicker. Don't delay and complain…get it done and move on. And find the interesting elements in the task – these are there so look for them. This in fact is the mark of a true leader as most people will not work well under conditions they do not enjoy. If you tackle it quickly, do your best and get beyond it you will find that it is best for you. A bonus from this is that you actually lead by example as others will notice. It will be silent among your colleagues but you will sense it and this will open an upward mobility path for you.

8.   CLOSING THOUGHTS

Starting your career is a great time in life. You are leaving school and most of your life will be spent working in an area you choose. You are leaving the dorm life, no more attending classes, nor regular attendance at school sporting events, no grabbing a bus to go out with friends....all great times for sure and will remain terrific memories. But now you need to turn your attention to the future and start on a new adventure. There are many factors to consider and writing them down helps to bring them to a conscious level. Even if one of the factors in your decision is to stay close to your parents when looking for a job, for example. Writing each of the items down provides a bird's eye view of what you consider important. If you do not bring them to a conscious level, they tend not to play a strong role in the final decision process.

Many careers are made as there are so many possibilities that exist in the workforce which you will see once you enter. Getting to know the company you work for, the market, the technology, the people, competitors, etc. all contribute to your choices for professional enhancement and development. In addition outside influences play a major role in decisions. Some of my colleagues have told me they stumbled into their role and enjoyed it but didn't really seek it. This can certainly happen but if you ask yourself what really motivates you and how that can fit into the marketplace you enjoy that will require some work and research. Most likely the role you really seek may not exist so creating it and demonstrating why it is of value to the company you work for (or want to work for) is exciting and rewarding. It is as though you are a pioneer out to construct the path you want to walk. It is not how you most likely will start but can be an option to consider as you climb your career ladder. This is what we are trained to do as scientists, mathematicians and engineers. We have an analytical mind so we can apply that to many problems. To find a career we need to parameterize the problem and that would consist of defining (but not limited to) the following:

1. What are your interests?
2. Does the job conflict with any of your values?
3. What do you enjoy talking about if the topic was of your choice (and it is…it is your job!)
4. What do your friends say you do well?
5. What accomplishments make you feel really proud?
6. What family issues do you need to consider?
7. Is the location important?
8. More interest in an office setting or out of the office (e.g. field application work or sales)
9. What is your desired salary range?
10. What will be your elevator speech?

These ten items as noted are the core questions to consider but certainly based on individuality you can find more. There will be a new set of parameters or answers to your questions each time you search for a job, so don't hesitate to review your answers. And always prioritize the list so you know your top three requirements. But once you have bounded the problem set out to research as much as you can about the employment sector of interest. Read job descriptions, talk to people, professional journals, corporate literature, social media, YouTube clips, etc. You will be working for more than thirty years once you enter the professional lifestyle so make it as happy as possible. So in summary:

- Have fun
- Remember interviewing is a game…some you win some you lose.
- Don't just evaluate based on a "checklist"….ask your heart if it is right for you.

And…..remember what the proton said….

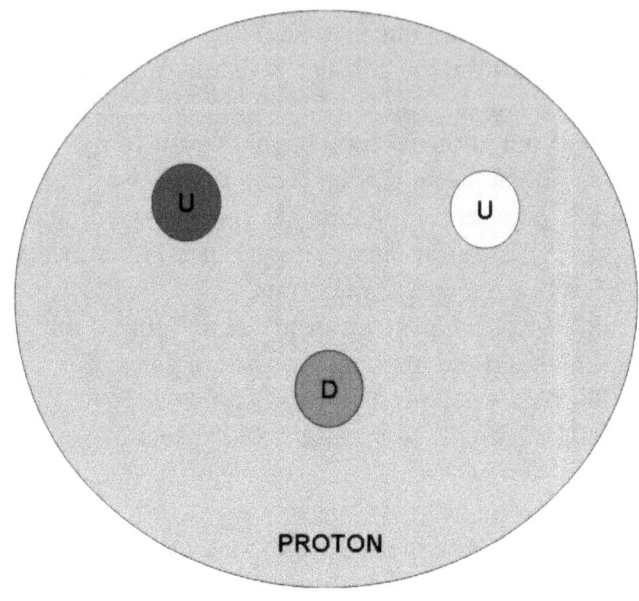

"Stay Positive!" ☺

Good Luck!

9. REFERENCES AND ADDITIONAL RESOURCES

1. COL, cost of living calculator,
    http://money.cnn.com/calculator/pf/cost-of-living/

2. Figure 1. Permission for use of this figure has been obtained and on record.

3. Figure 2. Permission for use of this figure has been obtained and on record.

4. Figures 8-15 are courtesy of AIP Statistical Research Center (www.aip.org/statistics).

5. Free Myers-Briggs Test.
    http://www.personalitypathways.com/type_inventory.html

6. Jung, Carl. *Psychological Types*. Important Books. 2013.

7. Mandiant APT1 Report.
    http://intelreport.mandiant.com/Mandiant_APT1_Report.pdf

8. Quenk, Naomi. Essentials of Myers-Briggs Type Indicator Assessment. Wiley. 2009.

9. Riley, Cindy. "Actuaries in action: Why it's rated the number one profession". STEM Education. February 23, 2013.

10. US Dept. of Labor Bureau of Labor Statistics. http://www.bls.gov/ooh.

Aushangpflichtige Gesetze
2016

Arbeitsschutzgesetz – ArbSchG; Jugendarbeitsschutzgesetz – JarbSchG; Arbeitszeitgesetz (ArbZG);Bundesurlaubsgesetz; Mutterschutzgesetz – MuSchG; Mindestlohngesetz – MiLoG; Kündigungsschutzgesetz (KschG); Allgemeines Gleichbehandlungsgesetz (AGG); Entgeltfortzahlungsgesetz; Bundeselterngeld- und Elternzeitgesetz – BEEG; Teilzeit- und Befristungsgesetz – TzBfG; Gesetz über Betriebsärzte, Sicherheitsingenieure und andere Fachkräfte für Arbeitssicherheit; AsiG; Gefahrstoffverordnung – GefStoffV; Röntgenverordnung – RöV; Heimarbeitsgesetz; Gesetz über den Ladenschluß – LadSchlG

Impressum

© MGJV-Verlag, Hans-Much-Weg 14, 20249 Hamburg, Telefon: 040/ 32030589;   Bitte wenden Sie sich mit Ihren Anliegen auch auf elektronischem Wege an uns: *MGJV.Verlag@gmail.com*

Inhaltliche Verantwortung und Aktualität: Redaktion MGJV

Satz, Druck und Bindung: Externe Dienstleister; alle Rechte MGJV-Verlag

Umschlaggestaltung: Gustav Broedecker. Alle Rechte MGJV-Verlag

Wir sind bemüht, ein ansprechendes Produkt zu gestalten, dass angemessenen Ansprüchen an das Preis/Leistungsverhältnis und vernünftigen Qualitätserwartungen gerecht wird. Konstruktive Anregungen nutzen wir gerne, um künftige Auflagen zu ergänzen und anzupassen.

- - - - -

Über eine Bewertung bei Amazon oder anderen Distributoren freut sich die Redaktion. Mit Kritik und Verbesserungsvorschlägen für künftige Ausgaben wenden Sie sich auch gerne an MGJV.Verlag@gmail.com

Vielen Dank, Ihre Redaktion MGJV

Aushangpflichtige Gesetze – Wozu sind sie notwendig und welche Verpflichtungen bestehen für den Arbeitgeber?

Durch Aushänge sollen die Arbeitnehmer über ihre Rechte informiert werden. Daher bestehen zahlreiche Vorschriften in unterschiedlichen Gesetzen, die Arbeitgeber verpflichten, eine Kenntnisnahme zu ermöglichen. Für die Arbeitgeber entsteht dadurch leider ein nicht unerheblicher administrativer Aufwand. Es empfiehlt sich daher, speziell zusammengestellte Gesetzes-Sammlungen wie die vorliegende an verschiedenen Stellen im Betrieb auszulegen. Nur so wird die angeordnete Fürsorgepflicht erfüllt und der Arbeitgeber vermeidet etwaige Schadensersatzansprüche und Geldbußen.

Je nach Regelung soll die Bekanntgabe soll in geeigneter Weise durch Auslegen, Aushängen oder Bekanntmachung geschehen. Für den Arbeitgeber ist eine hinreichende Lektüre der jeweiligen Vorschriften letztlich unvermeidbar, um die unterschiedlichen Vorgaben entsprechend umsetzen zu können.

Zudem muss für den Arbeitnehmer die Möglichkeit bestehen, sich ohne Schwierigkeiten über den aushangpflichten Inhalt zu informieren. Üblicherweise erfolgt ein Aushang an einem "schwarzen Brett" oder eine Auslage an einer allgemein zugänglichen Stelle des Betriebes. Häufig werden hierfür die Pausenräume genutzt.

Teilweise sehen die gesetzlichen Regelungen aber auch bestimmte Aushangsorte vor, zum Beispiel den Aushang der nach Heimarbeitergesetz erforderlichen Angaben in den Ausgaberäumen. Nicht ausreichend ist in häufig ein Hinterlegen bzw. Vorhalten im Personal- oder Lohnbüro. Bitte beachten Sie: Besteht ein Betriebsrat, so ist dieser über den Aushang zu unterrichten.

Sind von dem Aushang ausländische Mitarbeiter betroffen, die der deutschen Sprache nicht mächtig sind, kann das unter Umständen eine (jedenfalls zusammenfassende) Übersetzung erforderlich machen.

Bitte beachten Sie, dass keine Gesetzessammlung je Vollständigkeit garantieren kann. Eventuell sind branchenabhängig noch weitere Regelungen bekannt zu geben. Beachten Sie auch, dass Sie Tarifverträge und Betriebsvereinbarungen in der Regel in geeigneter Weise zugänglich machen müssen.